Children's Understanding of Mathematics: 11-16

The CSMS Mathematics Team

K. M. Hart   D. Kerslake
M. L. Brown   G. Ruddock
D. E. Küchemann   M. McCartney

Antony Rowe Publishing Services

This book has been printed digitally and produced in a standard specification in order to ensure its continuing availability

Published by Antony Rowe Publishing Services in 2004
2 Whittle Drive
Highfield Industrial Estate
Eastbourne
East Sussex
BN23 6QT
England

First published 1981 by John Murray (Publishers) Ltd

ISBN 1-905200-02-1

Printed and bound by Antony Rowe Ltd, Eastbourne

Foreword

This is an important book and I am honoured to have the opportunity of contributing a Foreword.

I may perhaps be said to have 'begun it all'. While running the original Nuffield Mathematics Project (mainly for children aged 5 to 11), I became aware of how much more we knew about the younger children than the secondary ones. The Nuffield project was able to draw (selectively and with caution) on the many experiments and 'check-ups' of Piaget and develop a hierarchy of concepts. This has been at the heart of the rethinking of mathematics in the primary school.

I have previously started a 'new' secondary scheme for abler children. Looking back, this was a very hit-and-miss affair (like all its contemporaries). The feeling grew that someone should look at the secondary scene and, as I put it (light-heartedly but not entirely without justification), 'do the difficult part in 5 years after Piaget had done the easy part in 40'. Finding just what secondary children are capable of learning, and where they are in their development, is indeed harder than at primary level, not only because previous research has been so scanty but also because the number of variables, and indeed the entire complexity of the problem, increases with age.

I discovered there were similar problems in Science and so got together with Professor Keohane and obtained a large grant from the Social Science Research Council. This book deals with the mathematical side of the subsequent work. My own self-appointed task was minimal, namely to harass the research workers and constantly demand my hierarchy or 'concept tree', from which authors and teachers could determine proper orders of topics and levels appropriate to the various children.

The brunt of my goading was borne by Dr. Kathy Hart. Although she has generously thanked the others in the Preface which follows, she has herself done a lion's share of the work.

The results reported below will have a shattering, but entirely beneficial, effect on the teaching of mathematics, not only in this country. In fact, everyone concerned with secondary mathematics teaching owes a debt to Dr. Hart and her collaborators, which can only be repaid by acting on their findings. In this way, mathematics can become more relevant, attainable and even friendly to future generations of school-children.

October 1979   GEOFFREY MATTHEWS
                                     Emeritus Professor of Mathematics Education
                                                           University of London

Preface

The research programme 'Concepts in Secondary Mathematics and Science' (CSMS) based at Chelsea College, University of London was funded by the Social Science Research Council for the years 1974–9. The research was firmly based in schools and on the school curriculum so that the topics investigated would be recognisable to teachers as the type of mathematics they taught.

This book describes the work of the mathematics team whose aim was to develop a hierarchy of understanding in mathematics which would provide information for teachers and other developers of curriculum. Eleven mathematical topics are discussed in detail and then levels within each are compared. The results of wide-scale testing are described in the book; the actual tests were published by NFER-Nelson.

The original directors of the programme were Professor Kevin Keohane and Professor Geoffrey Matthews, later when they were no longer at Chelsea College Professor Paul Black became director. We would like to acknowledge the help they gave us and also to thank Professors Eric Lunzer and David Johnson for their support.

At various times mathematics educators have joined the team to help write test items, interview children etc. Their help has always been welcome. We would particularly wish to express our appreciation to Tony Malpas (the first leader of the Mathematics team), Brian Joyce, Ian Stewart, Lesley Booth and Gabrielle Cornwell.

Finally our work would have been impossible without the active involvement of many teachers and children. We thank all those schools which volunteered their help and allowed us to test their children.

2004. Children's Understanding of Mathematics :11-16 has been reprinted nine times and this is the eleventh edition in the 23 years since it first appeared. We are now publishing with Antony Rowe Ltd. The test papers are available from Shell Publications, University of Nottingham. Kath Hart

Contents

1 The research of CSMS

Introduction

This book describes the hierarchies or levels of understanding in each topic for which test papers were written. Each chapter is devoted to a different topic and describes in each case the types of question children can answer easily and those which they find difficult. The methods they use to solve problems and the errors they make are discussed, together with the hierarchy in the topic. The chapters were written by different authors so may be in different styles but there is comparable information given on each topic. The methodology used in the research appears in this chapter and the statistics used together with the sample of children investigated appear in the Appendix. Chapter 13 gives an overview of all the hierarchies and the relationship between topic levels. The implications of the research for the teaching of mathematics are discussed in Chapter 14.

Throughout the book the children's ages are given. The approximate age levels used to describe the year in school are:

Top primary		– 11 years
1st year	⎫	– 12 years
2nd year	⎬ Secondary	– 13 years
3rd year		– 14 years
4th year	⎭	– 15 years

Teachers who may not wish to read all the book at one time should be able to find valuable information on particular mathematical topics by turning to the appropriate chapter.

Methodology

In order to present a picture of children's mathematical understanding which was representative of the English child population, it was necessary to use written tests. These were written by the CSMS team and were designed mainly in a problem-solving format in order to probe understanding rather than to test whether certain methods had been taught by a teacher. Items were written which were free of technical words and these were tried on interview with children and then replaced or revised. The interviews were used for a second purpose, that of finding the methods used and errors made by children when confronted with a mathematical problem. About thirty children of the appropriate age range, from different schools, were interviewed on each topic and their replies tape-recorded and later transcribed. The methods used by the pupils were often not 'teacher taught' but nor were they necessarily idiosyncratic as they were often repeated in different schools. The errors made could similarly be generalised and the answers obtained could later be observed in the written tests and interpreted with respect to the strategies which were likely to have led to them. The interviews were thus an integral part of the methodology, being used to inform and interpret the results of the written tests besides being used as a method of assessing the suitability of the test items.

The topics studied

(a)   Measurement (length, area, volume) for ages 12–14 +

(b)   Number operations (the meaning of the operations on whole numbers) for ages 11–12 +

(c)   Place value and decimals for ages 12–15 +

(d)   Fractions (fractional notation, addition and subtraction) for ages 12–13; extended to include multiplication and division for ages 14–15 +

(e)   Positive and negative numbers (notation, addition, subtraction and multiplication) for ages 13–15 +

(f)   Ratio and proportion for ages 13–15 +

(g)   Algebra (generalised arithmetic) for ages 13–15 +

(h)   Graphs for ages 13–15 +

(i)   Reflections and rotations for ages 13–15 +

(j)   Vectors and matrices for ages 14–15 +

Writing items

In the writing of items for each topic the writer of the test followed the following procedure:

(1)   Analyse the topic as it appears in commonly used mathematics text books and where relevant (e.g. ratio) as it appears in the science curriculum.

(2)   Investigate the results and examples used by other researchers with a view to using some of their findings.

(3)   Write a series of problems in the topic so that they span a wide range of difficulty, are free of technical terms and thought to be testing understanding rather than the repetition of a skill.

(4)   After team discussion, interview some thirty secondary school children covering the relevant ages and abilities, using the items as a basis for the interviews.

(5)   Modify or replace the test items, in the light of the interviews.

(6)   Using a class test try out the items in London secondary schools, the results being used to further modify items prior to the large scale testing.

(7)   Carry out large sample testing.

Interviews

Interviews took place in different schools within London and with children of different abilities. They lasted about an hour during which the child was asked to talk his way through the problems, explaining what he was doing at each stage. Not all children completed all the questions within the hour but it was thought preferable to have information in depth on a few items rather than superficial ideas on many. The interviews were tape-recorded and later transcribed.

A typical interview would include the child reading the items in order to see whether the questions involved the use of any word which the child found difficult or any word which could be termed 'technical' in that it was not used naturally and was known only if taught fairly recently. Such words were replaced when the items were revised. Considerable care was taken to make sure the interviewer understood the nature of the difficulty be it in the setting of the question or the actual mathematics. The first example below shows that the context of the question is confusing, and so it was rewritten prior to the final testing.

> In an office Mr. Adams comes in to work 2 days a week.
> Mr. Brown comes in to work 4 days a week.
> Mr. Carter comes in 6 days a week.
> The bill for making coffee in the office for these three men is 240p.
> How much should each pay for it to be fair?
> Mr. Adams . . . . Mr. Brown . . . . Mr. Carter . . . .

(I is the interviewer, J the child.)

J   10p, 8p, 6p.

I   Does that come to 240p?

J   That comes to 12 and I divided 12 into that which is 20, no that's wrong. 12 is the number of days they work a week.

I   Why did you bother to do that?

J   To tell you the honest truth I don't really know. Each one works 2 more days than the next one.

I   So who should pay the most?

J   Mr. Carter.

I   Any relationship between Adams and Brown?

J   Mr. Brown works 2 days more than Mr. Adams. It depends on how many times they have coffee a day.

I   What can we assume?

J   They have one cup a day. That has 2 cups of coffee a day, 4, 6, that makes 12 cups of coffee altogether in a week, 12 cups of coffee a day in a week.

I   Can you have that?

J   12 cups a week then and one each day. It must be 20p a cup of coffee.

I   You make the assumption they had 1 cup a day?

J   Oh yes, they might have 2 cups a day.

I   Would that make a difference?

J   Yes.

I   Do you think it would help if we knew how much a cup of coffee cost?

J   It wouldn't really help. He might have 3 and he might have 1.

The second example (below) demonstrates the efforts of the interviewer to examine exactly what the child understands and to make sure that the answer has not come about through carelessness. I — interviewer, ME — child.

Question: 10 sweets are shared between two boys so that one has 4 more than the other. How many does each get?

ME	How many *should* they get?
I	Yes, if one boy has 4 more than the other.
ME	One has 9 and one has 1.
I	Would one have 4 more than the other?
ME	Yes, because if you divide it by 2 boys, 10, then one would have 9 and the other would have to have 1.
I	I see. So which boy's got 4 more than the other?
ME	(Points to the '9')
I	Has he got 4 more than the other?
ME	Yes
I	How do you know he's got 4 more?
ME	Cos he's got 9 and the other's got 1
I	Does that make his 4 more?
ME	Yes
I	How?
ME	Cos I just worked it out. If there's 10, and divided by 2, so there's 5, and if that one's got 4 more, he's got 9, so the other has 1.
I	Why '1'?
ME	Cos there's only 1 left over.
I	And that boy now has 4 more than that boy?
ME	Yes
I	The one that's got 9 has got 4 more than the boy with 1?
ME	Yes.

Written tests

The test papers had to be so designed that they needed no specialised knowledge for administration. On some papers therefore, words commonly in use in that topic were defined, on others the children were given trial items and then the answers to these, as follows:

Algebra —— trial items demonstrating the use of the arrow in $x \rightarrow 3x$ and the use of a in $a + 4$, $4a$, were provided.

Reflections and rotations —— practice questions on reflections were given, the child was asked to check his answers by folding; rotation practice was given together with the answers.

Measurement —— the words perimeter, area and volume were defined with examples for the child (the answers were given).

Number operations —— a trial item was provided and discussed with the class for both the operations and stories. The children were reminded of the names and symbols for the four operations.

Vectors and matrices —— each term was introduced.

Place value and decimals —— the British standard notation used e.g. 30 000 rather than 30,000 and 4.29 rather than 4·29 was explained. A practice item explaining the naming system for the places with a picture to show the relationship between hundredths, tenths and a whole unit was provided.

Items of different levels of difficulty were written for each test but obviously the entire school syllabus in any topic could not be covered. Since the papers were to be given in the normal mathematics lesson, none could require more than an hour for completion.

Sample

The schools in the sample for testing were all volunteered by either the Head of Mathematics or the Headteacher. They were recruited by members of the team at in-service courses or meetings at Teacher Centres or sometimes from letters written to the team in response to reading an article about the research. The sample therefore came from schools where the teachers were sufficiently enthusiastic to attend courses and where they were fairly confident. The schools selected were predominantly outside London and were chosen in both urban and rural areas from as far south as Somerset and as far north as Manchester and Leeds.

The major periods of testing were in June—July of 1976 and 1977 i.e. at the end of the school year. The procedure for sampling in 1976 was slightly different from that in 1977. The general criteria for the selection of a sample were:

(1)   Each topic paper should be attempted by more than one year group (usually three year groups).

(2)   The sample for each year group on each topic should be representative of the English population of that age. In order to check on this a non-verbal I.Q. test was used which had previously been standardised on the English school population. The I.Q. distribution obtained from our sample for each year group in each topic was compared with the standardised normal distribution on I.Q. using the Kolmogorov-Smirnov one sample test or goodness-of-fit. (All second years in the schools from which the sample was drawn were given the NFER Test DH (non-verbal reasoning). Since the team could not find a suitable test of I.Q. for children over the age of 13, the other year groups in the school were assumed to have the same spread of I.Q. provided the teachers stated that the criteria for admission, the status of the school and the catchment area had not changed.)

(3)   For each age group entire year groups in 1976 and a quarter of each entire group in 1977 (a quarter of each class chosen by using random numbers) were tested. In 1977 all test papers were labelled with a child's name before distribution to schools.

(4)   For each topic, at least five schools were tested within each year group; the schools were from both urban and rural areas.

(5)   As far as possible each child did two test papers.

Each school was visited by a member of the team who spoke with the mathematics department and explained the aims of the research. On this visit the selection procedure for entry to the school was ascertained together with details of the class organisation e.g. mixed-ability teaching, streaming etc. Details of the sample used for each test appear in Appendix 1. Note that on the whole the children were from Comprehensive schools; some Grammar schools were used when the I.Q. distribution departed from the normal curve and extra children with high

I.Q. were needed. After the testing each school received the results for their children together with a general description of performance of all schools.

The size of the sample varied; in 1976 for example three thousand children took the Algebra test but that summer only four papers were being tested. In 1977 with twelve papers being used the samples for each topic tended to be smaller (the lowest number of children was 500). In all, fifty schools were used. A longitudinal survey in algebra, ratio and graphs was carried out over two years. Two hundred children were tested in each topic initially but after three testings (1976, 77, 78) there were about one hundred who had completed the topic test paper three times. The longitudinal study is reported in the chapter which compares the performance of different age groups.

In all, about ten thousand children were tested and their scripts returned to the CSMS office for marking. By coding the children's responses to each question it was possible to use a computer to determine both the facility of each item (i.e. the percentage of children giving the correct answer) and the frequencies of certain specific errors. Most of these wrong answers had been detected during the interviews but there were also some which were discovered during the marking. Some errors occurred in as much as 50 per cent of each year group. These are discussed in the context of the particular topic chapter together with a description of the methods used by children on interview which resulted in these same errors.

Administration of test

The teacher who adminstered the test was usually the teacher who taught that class mathematics and so she/he was asked to complete a questionnaire which asked: whether the topic had been taught that year; which items the children might find difficult; which books/schemes the teacher had used when teaching the topic. No attempt was made in the research to compare teaching schemes since it soon became apparent that most teachers used a variety of source materials in their presentation. A reply to the questionnaire which stated that only one text book was used was rare, because usually a text was supplemented by work sheets and work cards. Similarly no attempt was made to compare styles of teaching since the research team did not observe the teachers. It is recognised that the effect of a teacher on how the child learns is very important but it was impossible to control for this particular aspect in the research being reported. However, since so many children working with different teachers in different environments were tested it is thought that it should be possible to generalise the results to give a reasonably accurate picture of the levels of understanding in the English secondary school population. Indeed on the interviews it was found that children from very different educational backgrounds attending very different types of schools made the same type of error and often used the same methods.

The hierarchy

In many cases the order in which mathematics is presented to children is dictated by the needs of mathematics e.g. division before trigonometry or linear equations before quadratics. In other cases there is no clear order of presentation apparent in the mathematics and decisions have to be made by the teacher, based on experience and the dictates of the school syllabus. The research being reported took individual topics and attempted to form a hierarchy in each based on what the children tested appeared to understand. The data from the wide scale testing were used to form the hierarchy in each topic and therefore all the hierarchies are dependent on the items which appeared in the tests. Simply ordering the items according to facility (hard to easy) would make it difficult for the teacher to interpret the results and hence to decide the way the topic should be ordered. Therefore a method was sought by which items on a test paper could be grouped to form a 'type' and then the groups ordered to form a chain (easy to hard) in such a way that success on a harder group of items automatically entailed success on all easier groups. (The coefficients of association that were used are defined in Appendix 2.)

Items had to satisfy the following criteria before they were considered to form a group:
(a)   They should be of approximately the same level of difficulty i.e. same facility.
(b)   The values of the homogeneity coefficient ϕ (Appendix 2) item/item should be at an acceptable level.
(c)   The items should be linked (ϕ or Hij) with the items in both easier and harder groups.
(d)   There should be some measure of mathematical coherence to the items.
(e)   The groups should be scalable in the sense that a child's success (assessed as $\frac{2}{3}$ correct) on a group entailed success on all easier groups (Torgerson, 1958). (Error responses were not to exceed 7 per cent of the sample in any one test.)
(f)   There should be no gross discrepancies when each age group's results were analysed in the same way.

In some topics there was an obvious facility gap between clusters of items, which suggested a cut-off between groups. In other topics there was no obvious cut-off between groups and a decision was made on the basic of criteria (d) and (e) above. An acceptable level of ϕ depended on the test paper being considered; in some cases the easiest items were not highly correlated but were formed into a group in order that there should be some starting point to the hierarchy. The use of the criteria meant that several items on each test paper were rejected, usually because the correlation of a specific item with others of the same facility was noticeably lower. The values of ϕ were usually around 0.4 (1.00 showed perfect homogeneity); they are quoted in Appendix 2.

The groups of items obtained by the above methods were at different facilities and could be said to demand different 'levels of understanding'. The level of understanding displayed by a child in a particular topic could then be assessed by determining the most difficult group of items on which he could show competency. To do this a criterion of correctly answering two-thirds of the items within a group was used; where the number of items in a group was not divisible by 3 the whole number above or below $\frac{2}{3}$ was chosen, depending on which produced the better 'ordering'. A child who, in this sense 'passes' a harder group of items without passing all easier groups is said to be an 'error type'.

At the stage of matching each topic hierarchy to every other it was apparent that the facility range for each level on one topic did not exactly match the facility levels on a second topic; this is illustrated in Chapter 13. However, an approximate matching was obtained by dividing the facility range 0—100 per cent into four stages. The entire interconnection of hierarchies appears in Chapter 13 and should provide a useful guide to teachers who wish to compare levels of difficulty in one topic with those in another. A teacher who has dealt with Level 1 type problems in ratio may not wish to proceed to Level 2 in ratio (particularly as there is a large facility gap between the two). He or she may instead decide to start teaching algebra and would wish to know which was a comparable type of problem in that topic.

In order that a teacher may be able to assess the level of understanding of each child in the class (it will become apparent from the chapters on individual topics that there is a wide range of understanding in any one age group) the test papers developed by the CSMS mathematics team are available from NFER-Nelson, Windsor. Each is accompanied by a teachers' guide which gives details of the marking scheme and levels. In order to abide by copyright laws only about 50 per cent of the items on any paper are quoted in this book.

2 Measurement

The test paper dealing with measurement was designed to provide information on various basic aspects of the topic, particularly conservation. It was restricted to items testing the understanding of length, area and volume (and so excluded the measurement of time, angles etc.). The child was not required actually to do any measuring with a ruler but was asked to read the measures of lines from an already drawn scale. Many of the questions were in multiple choice format unlike the problems on other CSMS tests.

Topics covered

Piaget has carried out a considerable amount of research on the topic of measurement and many of the questions on the test paper were adapted from tasks he had used.

Length

The paper started with three questions adapted from Piaget, which dealt with the child's ability to compare the lengths of two line segments or a line segment and a curve. In two cases the end points of the segments being compared were aligned; this often proves to be a distractor and children who are not sure about the conservation of length are misled by the alignment into saying the segments are of equal length.

Measurement is taught in most British primary schools and an important aspect is the need for a standard unit. Children are often encouraged to use different units of measure prior to the introduction of a standard unit. The objectives are to show (1) that the number of units needed to describe the size of an object is dependent on the size of the measuring unit itself and (2) the convenience of using a standard unit. Both these aspects were incorporated into questions on length in the test paper. The ability to use a centimetre as a unit of measurement was tested by asking the child to state the length of line segments, a centimetre scale being drawn under the line segment, but in such a way that the line segment was not aligned with zero.

Area

The square unit to measure area can be used in two ways, either by covering a shape with square tiles or drawing a shape on squared paper and counting the number of squares covered. Both involve counting and not the use of a formula. A more difficult task is that of finding the area of a shape by counting squares and half squares or even estimating the approximate number when the shape has a curved boundary. The conservation of area was tested (i) by using an example where an equal number of holes were cut in two identical pieces of tin but the configuration of holes was different and (ii) by the presentation of a square which was then cut and rearranged in a different shape.

The formula for finding the area of a rectangle arises naturally from experiences involving the tiling of rectangles. Once the child has been taught the formula he is unfortunately likely to remember it in forms other than that which his teacher presented, e.g. $l + b$ instead of $l \times b$ or perimeter $(2l + 2b)$ instead of area. Examples where tiling was a convenient task were included, together with others where the use of a formula appeared to be necessary, e.g. find the shaded area

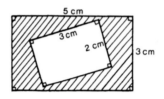

Perimeter

The 'rearrangement of area' tasks (where the area is conserved) were accompanied by questions on the nature of the new perimeter obtained. There is a powerful incentive to say the perimeter has not changed because the area has not changed.

Volume

Volume can be regarded as (1) the amount held by a container, (2) the number of units which when put together give the same configuration as the container (particularly when a number of cubes are compared with the space in a box) and (3) the displacement caused by putting an object into liquid. The last was omitted from the test paper but attempts were made to assess the understanding of the volume of a box and the conservation of volume when the cubes forming one figure were rearranged to form another.

Just as tiling can be used to find the area of a shape so the counting of cubes can be used when finding the volume of a solid. The introduction of examples where not all cubes are visible or where the dimensions involve half units makes the counting method rather laborious. Since the test was given in paper and pencil form the solids were shown in two-dimensional diagrams. On interview the children knew what the pictures represented. The only difficult diagram was that of a triangular prism when children found it hard to assess the dimensions of the triangular base from the two dimensional diagram, e.g.

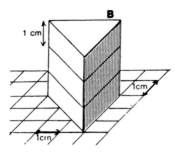

The children's responses

Length

A child is aware of the conservation of length if he realises that the number of units in a line segment does not change if the line segment is moved. Often children display a lack of awareness of the nature of measurement when presented with two unequal line segments (or a line segment and a curve) which have their end points aligned; they concentrate not on the length but on the end points. Piaget used examples like these prior to testing the equality of two sticks (the same length but one moved forward). The children's responses to the three questions concerning conservation of length are shown below.

1. The lines A, B, C, D, E, F are the dark lines on the squared paper below.
For each pair of lines, tick (√) the answer you think is true.

a)

	12 yrs	13 yrs	14 yrs	
(i) Line A is longer				
(ii) Line B is longer	86.4	90.0	93.0	per cent
(iii) A and B are the same length				
(iv) You cannot tell				

b)

	12 yrs	13 yrs	14 yrs	
(i) Line C is longer	42	45	52	per cent
(ii) Line D is longer				
(iii) C and D are the same length	48	48	45	per cent
(iv) You cannot tell				

c)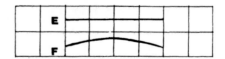

		12 yrs	13 yrs	14 yrs	
(i)	Line E is longer				
(ii)	Line F is longer	72	78	82	per cent
(iii)	E and F are the same length	19.5	15.5	13.4	per cent
(iv)	You cannot tell				

As can be seen many children judged the two lines segments in 1b to be equal; the fact that both were encased in four squares was seen by the children to be proof of equality. When interviewed most children who thought C and D were the same length gave their justification in terms of a number, e.g.

Child: They're the same. It doesn't matter what angle they are, just as
                long as they're 5 ...... 4 squares.
Interviewer: So you counted the squares?
Child: Yes, and if they end on, you know, the end, its the same.

Those who stated that C was longer than D did so in terms of movement:

Because if you pulled it down, it would go over the line.

The same mistake of stating two line segments are equal if they both go across a square is shown in Question 6 where again 40 per cent of the sample do not distinguish between the diagonal and the side of a square.

6. The 8-sided figure A is drawn below on *centimetre square* paper.

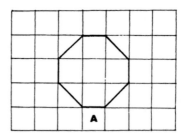

Draw a ring round the correct answer.
The distance all round the edge of A is:
8 cm   more than 8 cm   less than 8 cm   you cannot tell

Table 2.1   Responses to Question 6 (percentage)

Answer	12 yrs	13 yrs	14 yrs
8 cm	43.2	43.2	41.6
more than 8 cm	38.5	36.7	46.9
less than 8 cm	13.6	14.9	9.1

The use of units to measure length was tested in two contexts, the first with units of different length. One of the questions was as follows, in which parts a) and c) provided comparisons between numbers of the same unit, the difficult part is b) where the comparison is between numbers of different units.

3. John measures how long paths A and B are, using a walking stick. Then he measures how long paths C and D are, using a metal rod.
The answers are:

> Path A: 13 walking sticks
> Path B: 14½ walking sticks
> Path C: 15 rods
> Path D: 12½ rods

Draw a ring round the answer you think is true in each question:
- a) Path B is longer than Path A: True/False/Cannot Tell
- b) Path C is longer than Path B: True/False/Cannot Tell
- c) Path D is longer than Path C: True/False/Cannot Tell

Table 2.2   Responses to Question 3 (percentage)

	Answer		12 yrs	13 yrs	14 yrs
a)	Correct		91.1	93.7	96.2
b)	Path C is longer than Path B:	True	50.9	33.8	27.3
		False	7.7	7.7	6.2
		Cannot tell	37.9	56.8	65.4
c)	Correct		95.9	92.8	94.1

The interviews on this item showed again the child's dependence on a number answer, ignoring the unit of measure. The following conversation is taken from an interview on part b) of question 3 above where the child completely ignores the unit used to measure.

> Paul (aged 12.9) Yes, it's got more
> Interviewer What's got more?
> P C's got 15 and B's got 14½
> I C's got 15 what
> P Rods
> I What about B
> P It's got 14½. Sticks
> I So?
> P C's longest.

Many children when asked to use a ruler to measure a line segment count the units incorrectly, i.e. they very often start counting from 1 not 0. When the end of the line segment is aligned not with the end of the ruler but with some other point the error is compounded, for example:

5. How long is each line in centimetres (cm).

a)

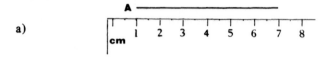

Length of line A:. . . . . .

Table 2.3   Responses to Question 5a (percentage)

Answer	12 yrs	13 yrs	14 yrs
6	49.1	64.9	76.1
7	46.2	30.6	22.8

The answer 7 could have been obtained by counting end points of the centimetres, not the gaps or by omitting to take note of the position of point A. The answer, as can be seen, is very common. A similar question where the line segment was measured from 6 to $10\frac{1}{4}$ was however successfully done by 66 per cent of the 12 year olds, the other two years scoring at very nearly the same rate as they did in 5a. There are between 10 and 18 per cent of each year however who count end points rather than spaces, given a line segment marked in centimetres when no distraction is present, e.g.

Area

The two questions dealing with conservation of area were adapted from questions used by Piaget and were as follows:

7. This picture shows two squares of tin which are the same size:

A machine makes 8 equal holes in each tin square:

A

B

Tick (√) the answer you think is true.
 1. Sheet A now has more tin . . . . .
 2. Sheet B now has more tin . . . . .

                   12 yrs 13 yrs 14 yrs

 3. A and B now have the same amount of tin 80   80   82 per cent
 4. You cannot tell if one now has more tin or not . . . . .
Give a reason for your answer: . . . .

8. I cut a square A into 3 pieces and arrange the pieces without overlapping to make a new shape B like this:

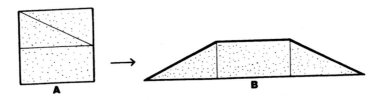

Tick (√) the answer you think is true in each question.
a) 1. A has the bigger AREA . . . . .
   2. B has the bigger AREA . . . . .

                  12 yrs 13 yrs 14 yrs

   3. A and B have equal AREA . . . . . 80.0   85.0   84.5 per cent
   4. You cannot tell if one AREA is bigger or not . . . . .

About 72 per cent of the total population could successfully answer both of these. The number of children who then went on to say the perimeters of the two figures in question 8 were the same (presumably because the areas were the same) was: 36 per cent (12); 29 per cent (13); 20 per cent (14).

The same confusion over equal areas and equal perimeters occurred in a question where the child was asked to draw a new rectangle on a given base such that its perimeter was the same as the original. Thirty-six per cent of each year drew a rectangle which had the same *area* as the original.

Finding the area of a rectangle by using square centimetres either for tiling or in a formula was successfully completed by 87 per cent of the total population. When the figure was composed of rectangles but was not itself a rectangle the success rate fell by 15 per cent. When the unit of measure was a small tile, ½ cm × ½ cm, 60 per cent of each year simply doubled the answer they had obtained when using a square centimetre, though both the small tile and the square centimetre were drawn on the paper. Counting whole and half squares to find an area was relatively easy (80 per cent success) but when the exercise involved matching quarter squares for example in question 13(b) the facility dropped to 57 per cent.

13(b).

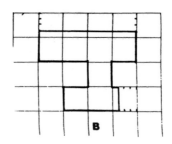

A much more difficult problem was the following:

13(c).   The area of the shaded figure measures 1 square centimetre. Find the area in square centimetres of shape C.

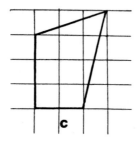

Table 2.4   Responses to Question 13(c) (percentage)

	12 yrs	13 yrs	14 yrs
Correct Answer	24.3	28.8	26.8
between 8 and 9	10.7	16.0	22.0
Answer 8	21.3	19.4	22.0
Answer 9	11.2	14.0	8.8

Finding the area by matching parts of a square to form a whole is difficult since none are halves; approximate matching would give the answers 8, 9 or 8 +. One method is to subtract two half rectangles from a rectangle (3 × 4), this requires the child to be familiar with the triangle-rectangle relationship and also to be confident enough to depart from a) formula, b) tiling, c) the confines of the figure.

Two questions provided a rectangle drawn on squared paper and a new base line. The child was asked to draw a rectangle on the new base equal in area to the original. This was achieved by some 60 per cent when the new height was an integer but only 20 per cent when the new height included a fraction. Fifty per cent of each year said the latter question was impossible to do. The original rectangle was drawn as below so the children knew that they were not restricted to the lines drawn on the grid.

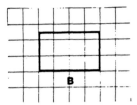

Volume

Piaget distinguishes between 'interior volume' where the space is confined within boundaries (e.g. a box) and 'occupied volume' where the volume in question is viewed in relation to other objects in the world around it. The latter is tested by presenting two different configurations of the same number of bricks. The CSMS version of this task appears below; the first two parts are simply an introduction, the rearrangement of blocks is tested in part c.

18.   A block 'A' is made by putting 8 small cubes like this together.

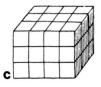

a) How many cubes make this block 'B' (there are no gaps inside)?. . . . .

b) Block 'C' is made by putting some small cubes together.

How many cubes make this block 'C' if there are no gaps inside? . . . .

c) All the cubes from block 'C' are put in a pile.

I am now going to use *all* these cubes from block 'C' to build a 'sky-scraper' so that the bottom floor is 4 cubes

How many cubes high would this 'sky-scraper' be from the ground? . . . .

The problem of finding the volume of a cuboid by counting cubes is that many cannot be seen. Even when the child is allowed to handle the cuboid those in the centre are of course still invisible. Children will often count what they can see and thus in 18b the answer 24 was given by 10.7 per cent of those aged 12, 13.5 per cent of those aged 13 and 12.6 per cent of those aged 14.

The correct answers for part c were at the 40–50 per cent level but a further 10 per cent of each provided an answer which was a quarter of the number they had given for the volume of the cuboid in part b. The volume of block C is of course obtainable by multiplication but some third year children on interview obtained their answer by addition, 'a layer of twelve and another layer and another'. Having obtained the answer 36 the last part can again be seen as an addition of fours until 36 is reached. Piaget states that the realisation that two volumes are equal if the products of their respective elements are the same is indicative of level IV (early formal).

The counting of cubes (even if done correctly) becomes very complicated when the dimensions are fractional, indeed the formula also becomes complicated since fractions have to be multiplied. The two examples below demonstrate this, there is a drop of over 40 per cent once the fractions are introduced.

20. The amount of room inside a block is called its VOLUME.

The VOLUME of block A measures 1 cubic centimetre

The VOLUME of block B measures $2\frac{1}{2}$ cubic centimetres

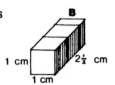

Find the VOLUME in cubic centimetres of each block C, D, E.

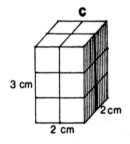

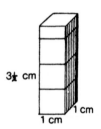

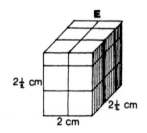

Table 2.5   Percentage of correct answers to Question 20

	12 yrs	13 yrs	14 yrs
Volume of C	56.2	57.0	68.1
Volume of D	65.7	70.7	81.5
Volume of E	14.2	18.7	27.9

The most difficult volume question was very similar to the comparable area question in that when the child was asked to use a different unit to measure volume he incorrectly stated the conversion rate between the two cubes used. Thus in the question below the child equates two small cubes to 1 centimetre cube.

19. How many 1 centimetre cubes would fit inside this box?

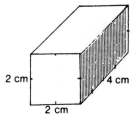

Number of 1 centimetre cubes: ....

b) How many small cubes like this would fit into the same box?

Number of small cubes: ....

Table 2.6   Responses to Question 19b (percentage)

Answer	12 yrs	13 yrs	14 yrs
128	0.6	8.3	6.2
32	71.6	61.3	63.3

Over 60 per cent of each year correctly answered questions on interior volume when a box had a cuboid of plasticine in it (how much space left?) and then when the plasticine was cut into three pieces and again placed in the box (how much space left?). This does leave 40 per cent of the secondary school age group who think that the remaining air space has changed simply because the form of the contents has changed.

Levels of understanding

The method for grouping items has been explained in Chapter 1. When this was applied to the measurement data it was found that some of the items adapted from Piaget's conservation tasks did not correlate highly with the other items. Also those items which involved the measurement of length were not sufficiently highly correlated to form a group at the easiest level of difficulty. In Table 2.7 the levels are described with the difficulty range; also the items involving 'length' and the Piagetian items are shown at their respective difficulty levels. Some items depend very much on the method used by the children for their solution. Thus finding the volume of a cuboid which has fractional dimensions is an easier task if the formula is used than if a counting method is attempted. The description of the levels in the hierarchy contain statements such as 'volume can be found by counting cubes' to distinguish between items where this seems a feasible method and those where the counting would prove to be difficult. The level four

Table 2.7   Hierarchy of levels in measurement

Level	Facility range (total sample)	Description of groups of items	Other items not in the groups at each level
0		Unable to do level 1 questions	
1	73%–89%	Area found by counting squares. Volume of a cuboid when only one cube on each layer.	Length questions 1a, 3a, 3c. Piaget conservation of area tasks.
2	58%–69%	Volume can be found by counting cubes when not all of them are shown. Simple application of area formula (examples of items at this level include 18b and plasticine in box item).	
3	40%–53%	Volume of cuboids when dimensions but not cubes are shown. Formula for area of rectangle needed. Area of a triangle (examples of items at this level include 18c).	
4	15%–24%	Application of area or volume formulae where the formula has to be adapted, e.g. half units or half formula are used. (Examples of items at this level include question 20E.)	
			Doubling area of a square (2 × 2)

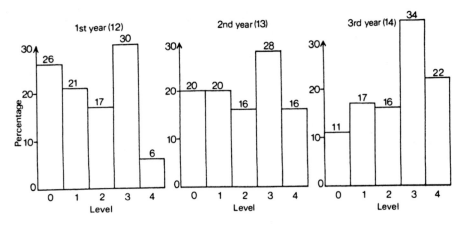

Fig. 2.1   Levels by year.

questions would seem to require the use of formulae and this might account for the difference in percentages of children at level four in the 4th year compared to the 2nd year.

Fig. 2.1 shows the percentage of each year at each level. There are less than 20 per cent of each year at level 2; this would seem to indicate that it is a transition stage. The description of level 2 items is very similar to that of level 3 items, the latter having slightly more complex applications. There are more 14 year olds than 12 year olds who can apply formulae and many more first years than third years who are unable to find the area by counting squares.

Implications for teaching

Measurement is one of those topics in mathematics which can be regarded as a prime candidate for the *maths is useful* description and indeed in the primary school considerable emphasis is placed on measurement skills. (We unfortunately have no information on the relation between practical measuring experiences and success on the measurement items.) It is apparent that the correct use of a ruler needs to be practised in the secondary school, which it often is not. When the results from this test paper were discussed with a group of London school teachers it was found that the only children (in their schools) who had measuring experience in mathematics lessons were in the remedial classes.

The words used in measurement activities are not common in everyday language, e.g. perimeter, volume. In fact they may have entirely different meanings in everyday language, e.g. from an interview:

Interviewer   Do you know what volume means?
Child   Yes
I   Could you explain to me what it means?
C   Yes, its what is on the knob on the television set

Many children being interviewed with the measurement paper in front of them often turned to the page on which the words were explained, to check what the question meant, showing that the technical measurement terms were *not* really part of their *normal vocabulary*.

As in other topics the results from the measurement paper show that the introduction of a *fraction* or any other complexity immediately makes an item very much more difficult. The results also show that children use naive methods, such as counting squares, for rather longer than one would expect and when these methods become cumbersome they are at a loss how to proceed. The formulae do not appear to be available; maybe they were taught when the child did not need them since he could solve area and volume problems by counting.

The essential aspects of measurement include:

(a)   The length, area and volume of objects are not changed by displacements.

(b)   Measurement can be quantified by the repetition of a unit of measurement but the resultant number depends on the size of the unit used.

(c)   Formulae for regular figures are short-cuts for the counting methods.

Perhaps 30 per cent of the secondary school children are not absolutely convinced of (a) above. About 50 per cent are not convinced that the line segments which form a side and diagonal of a rectangle are different in length. Some 40 per cent forget the importance of the unit used to measure, see the examples using metal rods and walking sticks to measure paths.

Thus there is clear evidence that the measurement skills acquired in the primary school need considerable reinforcement at the secondary level and indeed many children will not have acquired those skills at all and the problem is to teach them again. Other subject disciplines use aspects of measurement, for example Cookery, Science, Woodwork, and one wonders whether they rely on the mathematics department to provide the children with the basic knowledge and whether they experience the difficulties shown in mathematics.

3 Number operations

Introduction

This was the first of the CSMS investigations, and, being something of a pilot study, therefore differs in some respects from the investigations in other topic areas.

The aim was to try to examine children's understanding of the four basic number operations (addition, subtraction, multiplication and division) by finding to what extent children could both recognise which operation to apply in order to solve a 'word-problem' set in the 'real' world, and supply an appropriate context for a formal computation 'sum'. Initially the numbers in the problems were restricted to whole numbers, but problems involving decimals were later included as part of the 'decimals' test.

The area of 'application' of operations to real situations was selected in preference to the more obvious aspect of computation for the following reasons:

(i)   In a world in which pocket calculators are readily available there is likely to be a shift in emphasis in the mathematics curriculum away from laborious pencil and paper methods of computation and towards the selection of the correct buttons to press in a given problem.

(ii)   It is very difficult to disentangle understanding from rote procedure in computation; in particular children make a number of mistakes in computation which are rather specific to particular algorithms, and do not seem necessarily to indicate very much about the state of the child's general structure of understanding of that operation. (There is some discussion of this later in this chapter and also in Chapter 4 on place value and decimals).

(iii)   There are a number of straightforward and reasonably comprehensive tests of computation already on the market. In fact many of the children tested by us in this area were asked to work through such a test, and comments on the results are given later in this chapter.

We particularly wanted to investigate:

(a)   differences in difficulty between various 'models' of each of the operations $+, -, \times, \div$

(b)   the comparative difficulty of recognising the four operations $+, -, \times, \div$

(c)   the effect of the size of the numbers

(d)   the effect of using numbers as measures rather than for counting, i.e. whether the quantity was continuous or discrete

(e)   the difficulty of 'making up a story' (see item c below) compared with recognising the operation in a given story.

The selection of the items

Because there is evidence of children using 'key words' like 'share', 'times', 'altogether' to identify the operations an attempt was made to avoid the use of such words.

Five 'problem' items and one 'story' item from the final class test are given below.

1. A bar of chocolate can be broken into 12 squares.
There are 3 squares in a row.
How do you work out how many rows there are?

$$12 + 3 \qquad 3 \times 4 \qquad 12 \times 3 \qquad 3 - 12$$

$$6 + 6 \qquad 12 \div 3 \qquad 12 - 3 \qquad 3 \div 12$$

6. The Green family have to drive 261 miles to get from London to Leeds.
After driving 87 miles they stop for lunch.
How do you work out how far they still have to drive?

$$87 \times 3 \qquad 261 + 87 \qquad 87 \div 261 \qquad 261 - 87$$

$$261 \times 87 \qquad 261 \div 87 \qquad 87 - 261 \qquad 87 + 174$$

2. The signpost shows that it is 29 miles west to Grange and 58 miles east to Barton.
How do you work out how many miles it is from Grange to Barton?

$$29 + 58 \qquad 58 \div 29 \qquad 58 - 29 \qquad 29 \times 2$$
$$29 \div 58 \qquad 58 + 29 \qquad 58 \times 29 \qquad 87 - 29$$

7.   A shop makes sandwiches.
You can choose from 3 sorts of bread and 6 sorts of filling.
How do you work out how many different sandwiches you could choose?

$$3 \times 6 \qquad 6 - 3 \qquad 6 + 3 \qquad 3 - 6$$
$$18 \div 3 \qquad 6 \div 3 \qquad 6 \times 3 \qquad 3 + 3$$

8.   A gardener has 391 daffodils. These are to be planted in 23 flowerbeds.
Each flowerbed is to have the same number of daffodils.
How do you work out how many daffodils will be planted in each flowerbed?

$$391 - 23 \qquad 23 \div 391 \qquad 23 - 391 \qquad 391 \times 23$$
$$391 + 23 \qquad 23 + 23 \qquad 23 \times 17 \qquad 391 \div 23$$

c. $\boxed{9 \times 3}$ Story:

In the 'problem' items children were asked to ring only the *one* expression they thought 'best matched' the problem.

The testing

Sample

The results quoted are from testing different samples of children. Initially a large number of items were used but in the final test these were reduced in number; see Table 3.1.

Table 3.1   Sample

	No. of children	Age	Items
Interviews	35 (all ability)	12–13	Problems and stories
Initial Class Test	81 (biased towards lower ability)	12	20 problems, no stories
Final Class Test	497 (all ability)	11 ⎫	
Final Class Test	247 (all ability)	12 ⎬	9 problems, 5 stories
Final Class Test	130 (all ability but only two schools)	13 ⎭	

In all cases the questions were read out aloud by the teacher or researcher.

Children's ideas and strategies

(a)   *Various models of the four operations*

Not all real-life situations for which an expression like 8 × 4 forms a mathematical model are isomorphic.

Addition   The three most common models are:

(i) 'adding on' (e.g. I had five apples and I bought seven more; how many do I now have?) This is essentially unary in that an initial number is operated on.

$$+\,7$$
$$\text{⑤} \longrightarrow \text{②}$$

(ii) 'union' (e.g. I have five apples and you have seven apples; how many do we have altogether?) 'Union' is binary in that both numbers have a more equal status.

$$\text{⑤}$$
$$\quad\quad + \longrightarrow \text{②}$$
$$\text{⑦}$$

(iii) 'comparison' (e.g. I have five more than you. If you have seven how many do I have?)

$$?\xrightarrow{\ -5\ }7$$

There are also other variations (see for example item 2 on page 25 which illustrates 'directed lengths in opposite senses'), and the symbolic representations as shown above are not unique.

One difficulty is that word problems are sometimes not clearly categorisable as pure examples of only one model; they may contain elements of two or more models. For instance 'I had five apples. You brought your seven apples and put them with mine. How many are there now?' contains aspects of (i) and (ii).

In the initial class test (see Table 3.8 on page 40), the 'easy' model of addition $(+_1)$ chosen was 'union'. This turned out to be very simple to recognise, with 95 per cent of even this relatively low-ability sample getting the correct answer with large numbers.

The harder model $(+_2)$ involved 'directed lengths measured in opposite senses' as exemplified in the SIGNPOST item on page 25. (This was the only addition item to appear in the final test.) This proved very much harder than 'union' (with 73 per cent facility in the initial test). Some of the interviews suggested that it was the orientation which caused the difficulty, especially for the girls. For instance, given an earlier version of the problem which differed from the final one only in the numbers and units, Tracey (aged 11) asked:

Tracey Does it mean that is, er, 18 kilometres to Grange and 23 kilometres to, er Barton, does that mean that it's from the same place?

Interviewer That's right; from this signpost here. It's 18 miles that way to Grange and 23 miles that way to Barton.

T Take 18 away from 23 . . . 5.

Hilary (aged 11) was only able to do it when the situation was made more concrete.

Hilary Oh no, I'm no good at these . . . (pause) . . . you times those two together, don't you? . . . No, you can't . . . (long pause).

Interviewer Imagine standing there and you're looking up at the signpost, OK? Now that way it's 18 kilometres to Grange and that way it's 23 to Barton; we want to know the distance between the two.

H 23.

I 23?

H . . . (long pause) I'm not very good at doing kilometres . . .

I Let's try something else. We're sitting here, right? Say someone said it was three paces to the window and it was five paces to the window that way . . .

H You'd add them.

I . . . How far from one window to another?

H . . . (long pause) . . . eight.

I   Yes, what are you doing?

H   Adding them!

Of the children who on interview were able to make up stories for addition items, one third gave a 'union' model, e.g. Glenn (aged 12):

Note. Child's original spelling retained in stories.

3 men on a belDing Siet Tow of The men lade 84 Breks and one men lade 28 How would you work it out

A further third gave a straightforward 'adding on' example, e.g. Brian (aged 11):

John had 28p to go and spend and his mum gave him a nother 84p how much has he got now?

The remaining third gave a variant between 'comparison' and 'adding on' e.g. Juliet (aged 11) gave 'Karen has 9 eggs and Susan has 3 more'. In this latter case several children, like Juliet, had difficulty formulating the final question.

Of those unable to supply a story, some persistently misunderstood what was required. For example Russell (aged 12) tried for $9 + 3$: 'Two sixes', then '9 and 3 was walking down the road and 9 met 3 so they made 12' and even 'One day I was in class and I tried to get 9 add 3 is 12'.

Subtraction   For subtraction the three fairly common 'models' can be described as:

(i)   'taking away' (e.g. I had 12 apples and I removed five; how many were left?) 'Taking away' has a unary aspect in that you start with a number and then operate on it.

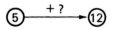

(ii)   'complementary addition' (e.g. How many apples do I have to add to five apples in order to obtain 12 apples?) 'Complementary addition' involves a kind of inverse of 'taking away'.

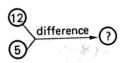

(iii)   'difference' (e.g. I have 12 apples and you have five apples, how many more do I have than you?) 'Differencing' has a more binary structure, starting off with two numbers and combining them.

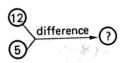

Variations other than these three are also possible,

e.g. $\quad$ (?) $\xrightarrow{\;+\,5\;}$ (12) $\qquad$ or $\qquad$ (12) $\diagdown$ difference $\longrightarrow$ (5) $\qquad$ and so on.
$\qquad\qquad\qquad\qquad\qquad\qquad\qquad\qquad$ (?)

The 'easier' examples ($-_1$) in the initial test included one 'take-away' ('DEBBIE') and two complementary additions, the one with large numbers being that quoted as item 6 on page 24.

On the initial test the 'take-away' was found very easy (86 per cent) compared with the small-number and large-number complementary additions (70 per cent and 44 per cent respectively). However on the final class tests (given to a representative sample) even the complementary additions had facilities for 12 year olds of 93 and 83 per cent respectively.

During the interviews it became clear that many children used an 'adding on' strategy to solve the complementary addition items. This meant that in some cases although they were able to use a strategy which was at least roughly correct, they did not recognise how the problem could be symbolised. For example Tony (aged 13) in answer to the GREENS item on page 24 (the numbers were 87 and 228 in this earlier version):

Tony	You add it on again . . . you add . . . three on to make 80(!) and then another 20 to make 100, then 128 from that is er . . . one hundred and forty something.
Interviewer	Which of those (expressions) do you think . . . ?
T	That one (87 + 228).
I	Are you sure it's that one? Did you add 87 onto 228?
T	No, I built it up.
I	You built it up? Do you think it's any of these (expressions)?
T	I think it's this one (87 ÷ 228).
I	87 divided by 228?
T	No . . . I don't know the sign for adding it on.

Others, like Anthony (aged 12) used this strategy but could also identify the problem as a 'subtraction' one, even if the formal subtraction algorithm was not remembered very clearly. (The numbers in the problem in this case were 78 and 204).

Anthony	Oh . . . take it away.
Interviewer	Take it away? Take what . . . ?
A	You take . . . well I don't do it like that, see, I sort of . . . I put two on that and that would be 80 . . . then you add 20 to 100 . . . that would be 22 . . . and add that 4 there, that would be 26 . . . and . . .
I	That's right. That's the best way of doing it really, isn't it? You don't have to write anything down. But if you had to write it down what would you have done with the two numbers there?

A	That one (selects 204 − 78).
I	Is that the only one?
A	That one up there (78 − 204).
I	That's a take-away. Is that the same as that one? Is it the same answer?
A	Yeh . . . no, course it won't . . .
I	Why not?
A	Or will it? 'Cos it's 78 . . . you put (writes 78

$$\begin{array}{r} 78 \\ 204 \\ \hline \end{array})$$

. . . put the four and two . . . and eight from four won't go . . . you have to borrow from there, and that makes 18 and that six. Four from 18 is 14 so you put down the four and put one there (carried). And then you go . . .

I	Sorry, what are you doing? Are you taking four from eight?
A	Yeh, but you have to go that way don't you?
I	You couldn't take four from eight?
A	No, no . . . oh yeh, that number has to be higher, don't it?
I	That has to be bigger, does it, than the eight? What would happen if you . . .
A	See, I sort of . . .
I	You can do it in your head better, can't you?
A	Yeh . . . yeh, that messes me all up, all that.

This then is an example of a boy who has quite a good feeling for number but who has not mastered the subtraction algorithm which seems to him to be based on an arbitrary set of rules. Many children like him who cannot successfully reproduce the algorithm may still be quite competent at solving problems set in a practical situation.

When children were asked to make up stories for subtraction sums they almost always gave a straight 'take away' model, e.g. Jane (aged 11):

Tony Had 84 Sweets He gave 28 to Fred How meny did he Have left

However there were occasional examples of complementary addition, e.g. Suzanne (aged 12):

A couple of people are on holiday for 84 day. They have been there 28 day already. How many more days does that leave them?

Those who didn't manage to produce a satisfactory answer either just got muddled or gave an addition problem.

Multiplication   Four of the common models are:

(i) 'multiplying factor' (e.g. I have seven apples and you have five times as many; how many do you have?) Again this is unary in flavour.

(ii)   'repeated addition' (e.g. I bought seven apples every day for five days; how many do I have altogether?)

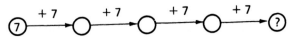

(iii)   'rate' (e.g. There were five people and each of them had seven apples; how many do they have between them.) This is very similar to (ii). It can be thought of as five people at a rate of seven apples per person.

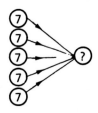

(iv)   'Cartesian (cross) product' (e.g. There are seven different types of apple, and each type is graded into five different sizes; how many different sorts of apple can you ask for?)
This is essentially a binary operation and can be represented best in a table (matrix) in which each entry is unique.

		A	B	C	D	E	F	G
	1	.	.	.	.	.	.	.
	2	.	.	.	.	.	.	.
Sizes	3	.	.	.	.	.	.	.
	4	.	.	.	.	.	.	.
	5	.	.	.	.	.	.	.

Types

The easier model chosen ($\times_1$) was 'repeated addition', and as expected this proved to be very much simpler than the 'Cartesian product' model (item 7, page 25).

Table 3.2   Comparison between Cartesian product and repeated addition − 12 yr olds final test (percentage)

Repeated addition		Cartesian product
(small numbers, continuous quantity)	(large numbers, discrete quantity)	(small numbers, discrete quantity)
87	77	62

In response to the 'repeated addition' item, many children did use an addition-based strategy, e.g. David (aged 12):

> David   Add up . . . by doing two 28's, and another two 28's, and do it again and again until you get to 19.

Interviewer That's fine. Do you think any of these (expressions) describes what you would do?

D (long pause) No.

A number of children indicated that they would use this strategy of arriving at the answer, but were able to verbalise it as '19 times 28', '28, 19 times', '29 times 19' or '19 28's', and thus to select either 19 × 28 or 28 × 19 as the correct expression.

The brighter children like Linda (aged 13) quickly suggested 'Multiply 28 by 19', or, in some cases 'Times 28 by 19', which suggested a multiplication, rather than a repeated addition, approach.

Many children, like Joseph (aged 12), found it difficult to understand the Cartesian product situation. The example below was CRISPS (four flavours and three sizes).

Interviewer (reads question, then long pause) . . . I'll explain this one . . . We have four different flavours, what flavour might we have?

Joseph Cheese and onion.

I So you'd have cheese and onion, three different sizes, so in one box we might have small cheese and onion, in another box we might have middle size cheese and onion packets . . . how many boxes would you need altogether?

J Three.

I Three for a small, medium . . .

J And big . . .

I All right, that's all right for cheese and onion, now there are four different flavours, each of them comes in three sizes. So how would you work out how many boxes there would be altogether? . . . (pause) . . . You said that for cheese and onion there'd be three boxes, that's right. Now there are four different flavours, cheese and onion's only one flavour, there are three other flavours, there's four altogether . . .

J Oh, there's four different flavours, one for each flavour.

I Right, but in each flavour we've got three different sizes.

J Oh, small, large and medium . . .

I Right, so for each . . . there are three boxes for each of the four flavours. How many boxes would you need altogether?

J Four.

I That would hold the four different flavours. Now for each flavour you've got three different sizes.

J Oh! . . . I know what you mean.

I Tell me how many boxes you'd need altogether?

J Oh, three! 'Cos there's three different sizes . . .

I That's for each . . . one flavour, but you'd have three for cheese and onion, then you might have three for plain . . .

J Oh! . . . I know what you mean . . . you mean four three's.

I That's right.

J I get that now, you ought to've explained that a bit more(!).

(J then selects 3 × 4 as the correct expression, and hesitates over 3 ÷ 4 also.)

Other children, after just such long explanations, arrived át 12 by adding four three's, and could not recognise the correct expression as 4 × 3 or 3 × 4.

When children were asked to make up their own problems, many chose either the 'repeated addition' model, or 'rate', or something between the two.

Shelley (aged 11) gave a very clear 'repeated addition' structure for 9 × 3:

> Julie bought 3 chews
> Jane bought 3 chews
> Mary bought 3 chews
> Andrew bought 3 chews
> Janet Bought 3 chews
> Shelley Bought 3 chews
> Margaret Bought 3 chews
> Peter Bought 3 chews
> Lee Bought 3 chews
> How many chews altogether

Jessie (aged 12) gave a 'rate' model:

> There are 84 skirts on each rack at a clothes shop.
> There are 28 racks. How many skirts altogether?

There were understandably very few 'Cartesian product' types but some, like Jeffrey (aged 11) gave a 'multiplying factor'.

> John had 9 times the amount of peters 3 marbles
> Who many did he have

Children who failed to produce a satisfactory story often ran into trouble through failure to select two quantities which would combine, e.g. Qamar (aged 12):

> Lee had 9 and Jim had 3 choolates If you mulity ply them how much do they have.

Division   The two most common models are:
(i)   'sharing' (or 'partition') (e.g. I had 35 apples to be shared between five people; how many each?)

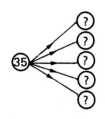

(ii)   'grouping' (or 'quotition') (e.g. I have 35 apples and I want to give five to each person; how many people can have a share?)

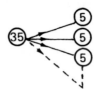

It can also be thought of as 'repeated subtraction'.

The first model ($\div_1$) was 'grouping' and seemed to be if anything a little harder than the second ($\div_2$) which was 'sharing' (see Table 3.3 below).

Table 3.3   Facilities of Division Items — 12 yr olds (percentage)

	Grouping	Sharing
Initial test — small numbers	63	83
Initial test — large numbers	53	56
Final test — large numbers	75	81

As in the multiplication example, there were several who on interview used a strategy of either repeated subtraction or repeated addition. For example Stephen (aged 12) on the DAFFODILS item (see page 25):

Interviewer	What would you do with those two numbers to work it out?
Stephen	23 . . . there . . . take away 23, 23, 23 . . .
I	Keep on taking away 23?
S	Yes.
I	Can you think of any of those (expressions) which would fit?
S	That one (23 ÷ 391).
I	Any others?
S	And that one (391 ÷ 23).

and David (aged 12), who on the parallel 'grouping' example added 13's to make 286 but could not find an expression which corresponded to what he was doing.

Both these strategies were correct, but depended only on a knowledge of addition or subtraction and not of division or multiplication. Again the brighter children answered immediately, e.g. Claire (aged 13): 'I think I'd go '23 into 391' but I don't know how much the answer would be.' (She then selected 391 ÷ 23 and rejected 23 ÷ 391.)

However this relatively bright girl, for whom none of the items presented any conceptual difficulties, did have trouble recalling the long division algorithm.

Claire   I'd do one of those things . . . (writes 23$\overline{)391}$) . . . but I don't think I know how to work it out like that. I've had it explained to me but . . . I'd have to do it the long way.

Nevertheless, she would obviously have been able to obtain the right answer by an informal strategy, and certainly would be able to use a calculator successfully.

With the small number grouping item, (CHOCOLATE, shown on page 24), the most usual response, both in interviews and in the class test, was to give the inverse operation 3×4 rather than the division symbolisation $12 \div 3$ (34 per cent of 12 yr olds in the final test gave 3×4, 32 per cent gave $12 \div 3$ and 13 per cent gave $3 \div 12$). For example Michael (aged 13):

Michael	Four.
Interviewer	Mm . . . how did you work that out?
M	Three times four.
I	Three times four is what?
M	12.
I	I see, how did you realise it was four?
M	Just worked it out in tables . . .
I	I see, OK. Which of these (expressions)?
M	Three fours (3×4).

M finally, when pushed, also accepts $12 \div 3$ but is then asked . . .

I	Which of those would you think the better one?
M	Three times four.

This showed that they were influenced by their method of solution rather than the structure of the question.

It was interesting that, although the 'sharing' and 'grouping' problems had proved roughly comparable in difficulty, in the story items almost all children produced a 'sharing' model, in many cases using the traditional 'sweets'. For example Colette (aged 12):

Karen had 84 apples and there were 28 of us all together how many did we each have

There were one or two children who used 'divided' or 'shared' in the sense given below (John (aged 12)):

Billy had 9 sweets and I had 3 So we dividithem

(b) *Comparison between operations*

It was clear from the interviews that addition was very well established as an operation in virtually all children. Indeed a few children on interviews used it as a 'universal strategy'. Maria (aged 13), for example, said (wrongly) 'you add them' and picked the addition expression for five consecutive problems before wondering 'Hey, how come they're all 'adds'?' Subtraction, in the 'taking away' form at least, was only slightly less well grasped.

However a small but significant number of children had very little concept of either multiplication or division, although they could in most cases have solved the problems using strategies which depended only on addition and subtraction.

It appeared that division was generally easier than multiplication because children could readily translate the ÷ into 'share', which gave them a concrete picture to hold onto.

Although in a straightforward multiplication problem children would often translate the situation into for example 19 28's, or 19 times 28, this did not make it very easy when children were asked to themselves construct a situation. The word 'times' itself did not readily suggest a concrete situation. In the long run this may be an advantage, as children may get stuck with the 'sharing' picture to the extent that they cannot conceive of division in cases where the answer is not a whole number (see Chapter 4).

The item facilities reported in Tables 3.8 and 3.9 also support the order of difficulty: +, −, ÷, ×. Thus in initial class tests, where the problem items for each operation were balanced as carefully as possible, the mean facilities were:

Table 3.4   Means and standard deviations of the facilities for the five problem items exemplifying each operation in the initial class tests (percentage)

	mean facility	standard deviation
+	87.6	9.5
−	67.0	15.9
÷	62.6	12.7
×	53.4	19.4

For the story items in the final class test the figures were:

Table 3.5   Facilities for the story-items in the final class test − 12 yr olds (percentage)

− (large numbers)	85
÷ (small numbers)	69
÷ (large numbers)	56
× (small numbers)	53
× (large numbers)	41

Alongside an overall order of difficulty +, −, ÷, ×, Tables 3.4 and 3.5 illustrate the considerable variation within each operation depending on the model used, the size of the numbers, and so on.

(c)   *Size of numbers involved*

Collis (1975a, b) had conjectured on the basis of his own results, which related to the 'structural' aspect of number, that children at the 'early concrete' level in Piagetian terms were able only to apply number operations successfully to 'small' numbers, whereas those at the 'late concrete' level were able to operate with numbers over a hundred. For this and other reasons it seemed worth comparing examples where the numbers were 'intuitable' i.e. under 12, with those where they were not, although in such cases we restricted ourselves to numbers between 20 and 500, and did not include more than one number over 100 in any question.

In the initial test were several pairs of problems using large and small numbers which were otherwise as similar in structure as possible; in the final test three such pairs were included (see Tables 3.8 and 3.9).

The small number items were always easier to recognise although the differences for parallel items ranged individually between 0 per cent and 27 per cent with a mean of about 14 per cent.

For the story items in the final test the differences between parallel small and large number items were again of the order of 15 per cent.

It does seem likely (and it was confirmed in interviews) that where the child was a little unsure of the operation with small numbers, the presence of large numbers pushed it over his or her threshold. Joseph (aged 12) expressed this feeling quite clearly.

Joseph	Um . . . You see I think they had, er . . . Oh! I don't know . . . I don't like all these big numbers.
Interviewer	The big numbers make it harder, do they?
J	Yes.

(d)   *Context — discrete or continuous quantities*

There seemed to be only one significant difference between different contexts in which the questions could be set, and this depended on whether the number related to a set of distinct objects (e.g. seven people) or whether it referred to continuous units of measurement (e.g. seven metres etc.). The differences in facility between the two types were consistently about 13 per cent, the continuous type being more difficult.

Almost all the stories given by the children had a discrete context, often using sweets, marbles etc. which suggests that the concrete way in which such operations were introduced in the primary school remains firmly fixed.

(e)   *'Problems' and 'stories'*

As has been seen, the story items were useful in both establishing whether children could produce a concrete situation to embody a symbolic expression and which 'models' were most prevalent. In almost all cases, not surprisingly, the stories were harder than even the most difficult 'problem' items.

The major difficulty in each case seemed to be that of choosing the units for each number, e.g. where 'dogs' and 'people' were chosen, addition became difficult, and where 'sweets' and 'sweets' were selected, addition was possible but not multiplication.

(f)   *Inversions and commutativity*

The readiness of children to identify expressions like $391 \div 23$ and $23 \div 391$ as being 'the same' was something which had not been expected at the beginning of the study. The exact number giving the inverse expression varied according to the context, the wording of the question and the order of presentation of the expressions, but the figures were as follows in the final test (12 yr olds).

Table 3.6   Facilities of corresponding correct and inverted expressions in final class test − 12 yr olds (percentage)

	Correct expression	Frequency of correct response	Frequency of inversion
Subtraction	8 − 2	68	17
	261 − 87	60	19
Division	12 ÷ 3	32	13
	286 ÷ 26	34	36
	391 ÷ 23	47	33

It will be seen from Table 3.6 that the tendency is less clear in the subtraction items, but in one of the division items more children give the inverted form than the correct one.

Table 3.7   Subtraction and division inversions (percentage)

	11 yrs	12 yrs	13 yrs
Subtraction (2 items)			
Both correct format	52	51	(54)
At least one inversion	22	27	(32)
Division (3 items)			
All three answers in correct format	19	17	(28)
Two answers in correct format,			
the third *not* inverted	9	9	(7)
One or more inversion	45	49	(49)

(Note: The figures for the 13 yr olds are not likely to be very reliable.)

The evidence from the interviews supports this. Ian (aged 12) gave a typical reply to one of the division items.

> Ian   391 divided *into* 17 (indicates 391 ÷ 17) and this one, 17 divided *by* 391 (indicates 17 ÷ 391).

This confusion between 'divided by' and 'divided into' is also typical; children tended to use them, along with 'shared between', quite interchangeably.

One got the impression that children felt that since, given two whole numbers and a division operation, there could be no doubt about the answer, it did not matter too much about the order in which they were written.

(g)   *Correlation between understanding and computation skills*

Of the children who completed the final class test, 557 also did an SRA diagnostic computation test, and 306 did the Staffordshire computation test (sheet 1 only).

The correlations were quite high at .69 and .60 respectively with the number operations test. However when one looked separately at the performance on multiplication and division on our test and the computation tests it was clear that some children had mastered the algorithms with only partial understanding of the operations. However those with hardly any understanding at all of the operations were not in general successful with the calculations. On the other hand there were some children with a high level of performance on the number operations test who performed poorly on the algorithms.

Results – The items

The results for the initial and for the final class-tests are given overleaf. The figures include those giving 'inversions' and 'inverse operations' and hence indicate recognition of the structure of the problem rather than its exact symbolisation.

Results – The children

The results of the final class-test were analysed in two ways.

(a)   *Assessing the operations separately*

Each child was assigned a 3-digit code (e.g. 201), each digit relating to performance on a group of items involving a particular operation. (For this purpose the one addition item was included in the subtraction group). A child could score one mark on each group if he answered correctly two out of three problems ('correct' answers included inversions and inverse operations as discussed earlier). In addition, a further mark was given for each group if the child could correctly supply a 'story' for the large-number examples using that operation. (The results for the small-number story items were ignored).

Thus if a child was assigned the 'code' 201, it would indicate that he or she had answered correctly at least two of the +/− problems *and* the (large-number) subtraction story, less than two of the multiplication items and failed on the large-number multiplication story, and either at least two of the division items, or, more rarely, the large-number division story, but not both.

Using this classification the percentage of children with the various codes for Years 0, 1 and 2 was as follows in Fig. 3.1. in which the code-numbers are arranged in order in a 3-dimensional lattice to show the progression.

The fact that the distribution of children is weighted towards the left side again illustrates the fact that division seems to be found an easier idea than multiplication.

Thus one can say that of children at the *end of the first year* in the secondary school, about 30 per cent will have a sound understanding of the number operations (code 222). This group on interview tended to give quick, confident answers and were able to use mathematical language in a fairly abstract way. (See, for example, Linda's response on page 32.)

The next 20 per cent (codes 212 and 221) are generally good at recognising multiplication and division in common contexts, but perhaps find it difficult to

Table 3.8 Percentage of successful answers (allowing inversions and inverse operations) to 'problem' items for initial class test (12 yr olds, N = 81, sample biased towards low ability)

	$+_1$	$+_2$	$-_1$	$-_2$	$\times_1$	$\times_2$	$\div_1$	$\div_2$
Sd	LESSONS 97		DEBBIE 86	POTATOES* 75	JANET 80	SHOP* 43	CHOC 67	TEAMS 83
Sc	MARKET 85	ALAN 88	JOHN 70		BUCKET* 67		TEACHER 78	
L	CRICKET 95	SIGNPOST 73	GREENS 44	JUG 60	CAKES 44	BUTTONS 33	PLAY 53	DAFFODILS 56

The abbreviations are as follows: S/L: small/large
c/d: continuous/discrete
→ :   indicates two parallel items which were designed to differ primarily in the size of numbers involved.
$+_1/+_2$:   'easy'/'hard' models of addition etc. respectively.

(It will be noted that some of the 'models' thought to be 'hard' in fact turned out relatively easy!)

*These items were re-written after the first trials and the percentage adjusted accordingly.

Table 3.9   Percentage of successful answers (allowing inversions and inverse operations) for final class test for 11, 12, 13 yr olds (representative samples)

	+	—	— story	x₁	x₂	× story	÷₁	÷₂	÷ story
Sd					SHOP 46 62 (74)	9 × 3 45 53 (54)	CHOC 71 78 (84)		9 ÷ 3 60 69 (62)
Sc		JOHN 91 93 (93)		BUCKETS 78 82 (82)					
L	SIGNPOST 74 78 (74)	GREENS 79 83 (91)	84 − 28 77 85 (89)	CAKES 66 77 (82)		84 × 28 31 41 (49)	PLAY 71 75 (85)	DAFFODILS 72 81 (82)	84 ÷ 28 42 56 (58)

Percentages are given in the order 11 yrs, 12 yrs, 13 yrs (N = 497, 247, 130 respectively). The 13 year old results are from two schools only, and should be taken as no more than a general indication of trend. Abbreviations are as in Table 3.8.

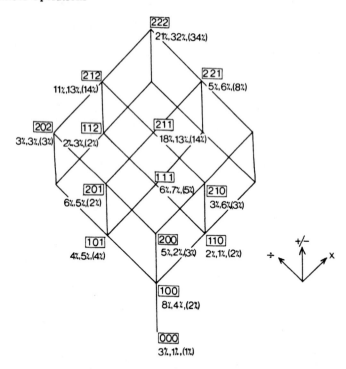

Fig. 3.1   Percentages of children aged 11, 12 and 13 years with each code number. (Data are not given for codes assigned to less than two per cent of each year.)

construct an appropriate 'story' when the numbers are large. (The weakness is about twice as likely to be on multiplication as division.) This group is more likely to be dependent on concrete strategies than the higher group, and is much more likely not to distinguish between the order of the numbers in the expression. (See for example Ian's response on page 38).

Of the remaining 50 per cent, up to 10 per cent at the lower end (codes 000, 100, 200) have very little concept of the operations of multiplication and division, although as we have seen they may well be able to use addition-based strategies to solve concrete multiplication and division problems. (The responses of David quoted on pages 31 and 34 exemplify this group.) The remaining 40 per cent seem likely to be groping towards ideas of multiplication and division, and are at various stages on the way. (Joseph quoted on page 32 would probably come into this group.)

(b)   *Using a hierarchy*

In the second method all items, with the exception of the addition item which did not correlate very well with the others, were grouped into three levels of difficulty, in the same way as for other tests. Because of the small number of items, the ϕ-values were lower than normal, and it was only possible to create three levels, the hardest covering a very wide facility band.

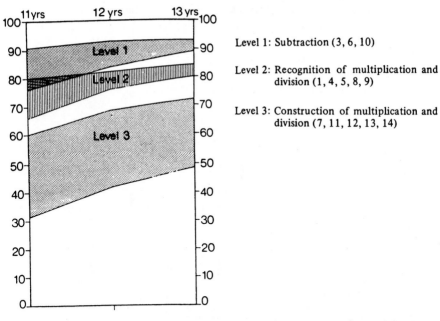

Level 1: Subtraction (3, 6, 10)

Level 2: Recognition of multiplication and division (1, 4, 5, 8, 9)

Level 3: Construction of multiplication and division (7, 11, 12, 13, 14)

Fig. 3.2   Levels for 'Number Operations' showing facility bands for each level.

Level 1 contains all the subtraction items (again inversions and inverse operations are included as correct answers); Level 2 contains all the multiplication and division 'problem' items with the exception of the Cartesian product multiplication; Level 3 contains the Cartesian product multiplication problem and all four multiplication and division 'story' items.

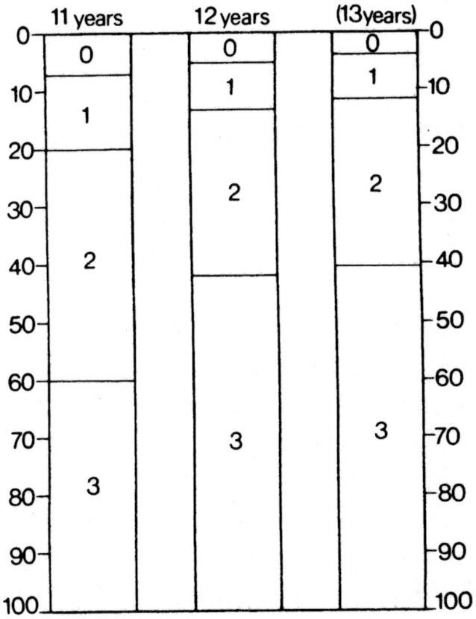

Fig. 3.3   Proportion of children from each year at each level.

Assigning children to levels as in the procedure described in Chapter 1, we have the comparison of levels in each year group in Fig. 3.3.

Again we obtain a figure of around 50 per cent of children at the end of the first year in the secondary school with a reasonable grasp of multiplication and division, and a lowest group of around ten per cent who at best can cope with addition and subtraction, with the remainder in between.

Decimals

Items concerning the recognition of the correct operation corresponding to a problem were also incorporated into the class-test on 'Place value and decimals', e.g.

19.   Ring the CALCULATION you would need to do to find the answer:
A.   A table is 92.3 centimetres long. About how many inches is this?
(1 inch is about 2.54 centimetres.)

$$2.54 + 92.3 \qquad 92.3 \div 2.54 \qquad 2.54 \div 92.3 \qquad 92.3 - 2.54$$

$$2.54 - 92.3 \qquad 92.3 \times 2.54$$

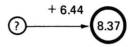

(two other parts of item 19 are quoted on p. 55)

(a)   *Different models of each operation*

There is again a variety of types of models, although some of the models involving whole numbers (e.g. 'repeated addition') are no longer possible and others become more common. In *addition and subtraction* the models are similar to those listed previously.

The only subtraction model included in the test was that which can be represented as

$$+ 6.44$$

$$\text{?} \longrightarrow \text{8.37}$$

For *multiplication*, the possible models appeared to include 'multiplying factor', 'rate' and 'Cartesian product'. ('Cartesian product' is similar to the idea of calculation of area, since both are symmetric and suggest rectangular arrays.) The most common model is probably 'rate', in which a rate is given together with a quantity (in this case one of the examples involved a rate of cost per kilogram and the number of kilograms, asking for the total cost to be calculated). Both examples in the test involved the 'rate' model, and the major difference between them lay in the fact that one contained a number less than one; this depressed the facility of that item significantly from between 14 per cent to 33 per cent according to age-group.

For *division*, the sharing ('partition') model generalises into one where a rate is demanded, given two quantities. For, in a 'sharing' situation, one is asked to find the 'number per person', given the total quantity and the number of people; this might generalise for instance into finding the cost per gallon given the total cost and the quantity of petrol in gallons.

On the other hand the 'grouping' (quotition) model generalises into finding one of the quantities given the rate and the other quantity. One moves from being given the total number, and the 'number per person', and being required to find the number of people, to a situation such as that in the example on page 44 where one is given the length in centimetres, the exchange rate i.e. the number of centimetres per inch, and one is asked to find the number of inches.

One example of each type was included in the test. However the comparison between the two types was complicated in practice by the fact that in one of them only ($\div_B$, the 'sharing' model) it was necessary to divide the smaller number by the larger one. If one therefore adds in the inversions, there is a tendency for the example when the rate was given ($\div_A$, the 'grouping' item shown on page 44) to be slightly harder to identify than the other, as was the case with whole numbers.

Table 3.10   Percentages of children aged 12, 13, 14 and 15 years giving correct answers to problem items involving decimals

	—	$\times_A$	$\times_B$	$\div_A$	$\div_B$
Age 12	52	32	18	27	19
(N = 170)	(63)			(44)	(53)
Age 13	52	42	17	35	23
(N = 294)	(67)			(50)	(62)
Age 14	68	54	21	44	30
(N = 247)	(76)			(63)	(73)
Age 15	63	53	29	45	28
(N = 238)	(69)			(57)	(68)

The numbers in brackets indicate the percentages if the inverted expression is also counted as 'correct'.

(b)   *Comparison between decimals and whole numbers*

Fig. 3.4 illustrates the fact that the presence of decimals in a problem makes it much harder to identify what operation is needed.

Provided inversions are again allowed, the order of difficulty of operations $+, -, \div, \times$ is maintained.

(c)   *Inversions and commutativity*

The results for subtraction in Table 3.11 are consistent with those on page 38 for whole numbers.

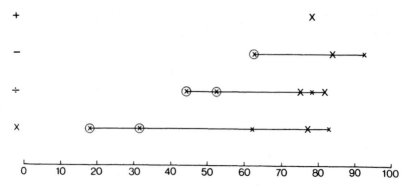

Fig. 3.4 Percentage of 12 year olds who gave correct answers (including inversions and inverse operations) to problem items involving small numbers (x), large numbers (X), and decimals (⊗).

Table 3.11 The incidence of inversions in the items involving decimals (percentage)

	Correct expression	Frequency of correct answer (Ages 12, 13, 14, 15)	Frequency of inverted answer (Ages 12, 13, 14, 15)
Subtraction	8.37 − 6.44	52, 52, 68, 63	11, 15, 8, 6
Division	92.3 ÷ 2.54	27, 35, 44, 45	17, 15, 19, 12
	4.86 ÷ 6.44	19, 23, 30, 28	34, 39, 43, 40

The figures for division suggest the following:

Table 3.12 Knowledge of correct order of division expression at ages 12 and 15 years

	12 years	15 years
Order consistently correct (decimals)	under 5 per cent	under 10 per cent
Order consistently correct (whole numbers)	20 to 30 per cent	(unknown)

This is supported by interview evidence as only one child out of 39 in the 12–15 age range gave a correct reason for choosing 4.86 ÷ 6.44. The remainder either used a 'large number first' rule or did not distinguish between 4.86 ÷ 6.44 and 6.44 ÷ 4.86 and hence chose randomly.

Implications for teaching

In a situation where calculators are readily available for computation, it is clear that the emphasis must change from algorithm-learning to understanding of the structure of the operations themselves and how and when they should be applied.

However with 10 per cent of children at the end of the first year in secondary school having little appreciation at all of multiplication and division, and another 40 per cent being shaky on these concepts even where only whole numbers are involved, there is little room for complacency. Secondary teachers may mistakenly take it for granted that such ideas are necessarily picked up at primary level.

Although children could to varying extents recognise situations other than those embodying the most common models of the operations (i.e. take away, sharing, repeated addition) it was clear from the responses to the story items that these forms were firmly implanted. Perhaps more should be done in both primary and secondary school to investigate alternative 'models' of applications. The 'Cartesian product' model of multiplication in particular was one which would merit more attention in both the secondary and primary classroom. More practice in supplying a variety of 'stories' for a single expression would help here.

While most children appeared to appreciate that $4 + 3$ and $3 + 4$, and 4×3 and 3×4, were equivalent, only between 20 per cent and 30 per cent of the 12 yr olds recognised the non-commutativity of division (with rather more in the case of subtraction). While this may not be in itself very important it will lead to incorrect results with calculators. It is also likely to lead to problems over the introduction of negative numbers (where $3 - 7$ must be distinguished from $7 - 3$) and fractions (where $2 \div 8$ is not the same as $8 \div 2$).

On interviews it was clear that many children, and even those at higher levels, resorted to informal addition-based strategies when asked to solve practical problems. In a number of cases (some examples are given) this was due to inability to recall the standard algorithms. It may well be the case that a combination of reliable mental methods and the ability to use a calculator are sufficient for all practical purposes. If teachers do feel it worthwhile to teach pencil and paper algorithms, then either more time must be devoted to practising and recalling them, or they must be better related to children's knowledge to assist recall. Perhaps the present methods should be abandoned in favour of others, maybe less efficient, but more related to children's own informal methods, and hence easier to remember. (See also Chapter 4.)

Where decimals were included in the problems, children found it very much more difficult to select the correct expression, especially in some of the particular awkward cases (e.g. multiplying by a number less than one, dividing a smaller by a larger number) where correct response rates dropped to under 20 per cent. In particular hardly any 12 yr olds and under 10 per cent of 15 yr olds would consistently be able to press the buttons on their calculator in the correct order in solving simple division problems. Hence much more work needs to be done before children are even able to use a calculator efficiently to solve straightforward problems which occur in real life.

4 Place value and decimals

Introduction

The aim of this area of the CSMS work was to find whether children could meaningfully use the base-ten place-value notation for both whole numbers and decimals, in the sense of both understanding how it worked and applying it to appropriate situations. The study concentrated on the area of decimals since most children in the 11–15 year age group were expected to have a reasonably sound basic knowledge of whole numbers.

There was a major problem in trying to differentiate between the 'relational' understanding of properties of number, and a mere 'instrumental' ability to carry out a technique. (In fact a few formal arithmetical computation items were included for comparison purposes.)

It was decided that as little explicit reference as possible would be made to fractions, although clearly the decimal system rests on the basis of this idea. In particular no fractional notation was used (.3 was described as either 'three tenths' or '3 tenths', but never as $\frac{3}{10}$), and there were no items requiring direct conversion between the two systems, although several questions required an implicit correspondence to be made.

Little reference was made to money in the test. This was because the decimal system of money is atypical of decimal systems in general (e.g. £6.4 has no clear meaning and £0.42 is interpreted as forty-two pence rather than as a fraction of a pound).

Topics covered

The major ideas which were incorporated are listed below. Examples which help to illustrate the meaning of these descriptions are given in the following sections.
(a) Correspondence between 'name' and 'place'.
(b) The 'carrying' aspect of addition, i.e. ten in one place is equivalent to one in the next place on the left.
(c) The 'borrowing' aspect of subtraction which is the reverse of (b) above, i.e. one in one place is equivalent to ten of the next place on the right.
(d) Other relations between places, including comparison of size of two numbers.
(e) Visual correspondence with fractions, both by length and area, as used in reading scales.
(f) Significant figures and approximation.

48

(g)   The effect of multiplication by a multiple of a power of 10.
(h)   The effect of division by a multiple of a power of 10.
(i)   The notion that multiplication by a number less than one decreases the initial number, while division by such a number increases it.
(j)   The infinite nature of the set of real numbers.
(k)   Knowledge of the type of real situations in which decimals are normally used.

It was intended to include a section to investigate understanding of the place-value system by using concrete examples involving bases other than ten, but because of the time that it would have taken to explain such questions this regretfully had to be omitted.

Children's ideas and strategies

Whole numbers − place value

The weakest group of children, when interviewed, often showed a superficial knowledge of place-names which could have easily led one to conclude that they understood the ideas of place value. It usually required a little more probing in order to expose the shakiness of their grasp.

Jack (aged 13, second year) was a typical example. He was a mature and articulate middle-band boy who was thought by his teacher to be 'about average' for his age-group. But although he could read the number 521 400 correctly, 8030 was read as 'eight hundred and thirty'. Similarly he rapidly calculated the correct oral answer to 'add ten to 3597' and 'add one hundred to 19 930', but happily recorded them as 367 and 2030, respectively. When doing a standard subtraction sum he 'borrowed tens' quite indiscriminately; from the thousands column over to the units, then from the thousands to the hundreds, and finally from the units back to the thousands!

One question which acted as a useful test of children's grasp of place value is that below.

7.   This meter counts the people going into a football stand.

$$\boxed{0\,|\,6\,|\,3\,|\,9\,|\,9}$$

After *one* more person has gone in, it will read:

$$\boxed{||||}$$

Age	12	13	14	15
Facility	68	77	86	88 per cent

Shakeel (aged 12, second year), who later turned out to be capable of performing long multiplication of decimals, gave $\boxed{0\,|\,6\,|\,3\,|\,1\,|\,00}$ since 'one to ninety-nine gives a hundred'. Raymond (aged 12, second year) first tried $\boxed{0\,|\,6\,|\,3\,|\,9\,|\,9}\,1$ and after the interviewer had explained that there were only the given number of 'holes', altered three to four thus: | 0 | 6 | 4 | 9 | 9 | . He explained 'It can't go there (points to 9 in 'units') cos it makes 10, and it can't go there (points to 9 in 'tens') as it makes 10, so . . .'

Raymond had however earlier managed to successfully 'add ten to 3597', and to 'add two hundred to 19 930', doing both operations in his head. But when asked to 'take away two hundred from 3104' he arrived at 3003 explaining that because he could only take away one of the hundreds, he had taken the other one off the 4 (units).

Again we have an example of a tenuous grasp of place value which breaks down when the examples seem to present particular difficulties.

One or two questions concerned with place value, especially those containing numbers over a thousand, exposed uncertainties even among children of average attainment. The first item below is one concerned with the use of zeros as place-holders, and the second involves the apparently straightforward naming of places.

3. Write in figures;
    Four hundred thousand and seventy three.

Age	12	13	14	15
Facility	42	51	57	57 per cent

1b. 5214   The 2 stands for 2 *HUNDREDS*
     521 400   The 2 stands for 2 . . . . . .

Age	12	13	14	15
Facility	22	32	31	43 per cent

Whole numbers – routine computation

It was clear from the written test results that many pupils had difficulty with a straightforward computation item in subtraction.

16b. *Subtract* 2312
                 − 547
                 ‾‾‾‾

Age	12	13	14	15
Facility	61	61	62	66 per cent

(A similar addition item with two three-digit numbers with carrying in each place was done correctly by 85, 89, 88 and 88 per cent of children aged 12 to 15 respectively.)

The interviews threw some light on these difficulties. Some children were just unsure of 'the rules'. Maria (aged 13, third year) is a good example of a pupil who can obtain the right answer once she is reminded of one critical step. In this interview the actual computation given was

          51
        −28
        ‾‾‾

Maria   Do you take the top from the bottom? (Tries, and takes one from eight, writing seven in the answer).
Can't take five from two — have to take one of these (indicates 1 in 51).
Interviewer Explain what you did there.
M   Crossed out the one (of 51) and put nought (in it's place) and put the one on there (i.e. to the left of five to make 15).
I   What was that to do? Why did you do that?
M   I put one on the tens.
I   O.K. Right, now what are you going to do?
M   That's wrong.
I   Why is it wrong?
M   I'm supposed to take 15 from two and not two from 15 (pause)
I   Can we do it the other way? Can we do it this time so that we take the bottom one from the top one?
M   Is that how we're supposed to do it?
I   That's how you usually do it, yes.

(Maria now proceeded correctly, and explained clearly how the 'borrowing' of the ten works.)

Some children who would have been assessed as lacking any real understanding of place-value obtained correct answers with correct, though stereotyped, explanations. However other children who showed considerable grasp of place value and decimals were unable to do so. For instance Peter (aged 13, 3rd year) was assessed in the interview as average for his year group, but obtained 37 for the answer to   51 and was unable to give any more explanation than
   −28·
   ‾‾

   ‾‾

. . . 1 take away 8 is 7, 5 take away 2 is 3 . . .

This failure to relate 'the rules' to the, perhaps considerable, understanding that children may have of the place-value system, may well account for the lack of correlation between this item and most of the others, and for the failure of the facility of this question to increase over the four years at the same rate as do the facilities of most of the other questions.

Decimals — basic 'values' of places (tenths, hundredths etc.)

The major difficulty that weaker children seemed to have was in understanding that the figures after the point indicated that part of the number which was less than one unit, even though the names of the decimal places in relation to diagrams showing units, tenths and hundredths had been explained to them at the beginning of the interview or test.

Instead, children seemed to think that the figures after the point represented a 'different' number which also had tens, units etc.

This is illustrated by the following question and a typical response:

10a.   Ring the BIGGER of the two numbers: 0.75 or 0.8. Why is it bigger?

Age	12	13	14	15
Facility	57	65	69	75 per cent

Jane (aged 12, 2nd year) ringed 0.75 because:

This is nothing before and seventy-five; this is nothing before and just eight.

When pupils were asked to write down how they would *say* the number 0.29 the percentages were:

Table 4.1   Verbalisation of 0.29

	12 yrs	13 yrs	14 yrs	15 yrs
(Nought) point two nine	26	32	41	41
(Nought) point twenty-nine	25	32	30	27
twenty-nine	19	13	8	10

Decimals – relations between places

There were a number of children who in the interviews showed that they were at the stage of trying to resolve some of the relationships between decimal places, and their answers therefore showed some inconsistencies. Among these were Frances (11, 1st year), Julie and Cecilia (12, 2nd year), Billy (14, 3rd year) and Kevin (14, 4th year). They were each busy fitting together the jigsaw, but often they had different pieces completed. At least they seemed to share the belief that there was a logical structure underlying it all.

When asked to 'add one tenth to 2.9' only Frances produced the correct answer. Julie, Cecilia and Kevin got 2.19 ('add ten to the tenths and you'll get 19') whereas Billy arrived at 2.10 ('add one onto 9').

However, surprisingly, all except Billy managed to correctly write down 'eleven tenths' as 1.1 ('ten tenths is one, and one over', 'cos 11 is over 10 so I think it'd be there as it's gone over'). But only Frances and Cecilia managed to extend their correct reasoning on 'eleven tenths' to 'eleven hundredths', the others mostly writing down .011.

Frances had given as the answer to 'Is there a difference between 4.90 and 4.9?' 'Yes, 4.90 is more', but in the next question decided that 0.8 was bigger than 0.75 because: 'Oh, it is eight tenths which equals 80 hundredths'. This was said with great satisfaction, as if somehow it had all come together.

When asked to 'multiply by ten the number 5.13', none of this group resorted to a long multiplication algorithm. Frances and Kevin appeared quite happy with 5.130 ('. . . you add a nought on the end – my Dad told me about that'). Julie and Cecilia each arrived at 50.130 ('cos ten times five is 50, so you just add a nought on the end'). They both seemed to obtain initially 5.130 using the rule (Cecilia had actually written this down), then rejected it as not being ten times as great as five, and hence tried modifying the rule to give an answer which appeared about the right size.

Billy reasoned his way right to the end, given a small hint.

> Billy   ... (pause) ... You can't put a nought on the end of there as it's a decimal ... (long pause) ...
>
> Interviewer   How big, roughly ... ?
>
> B   50 ... 51.3
>
> I   How?
>
> B   I multiplied that (5) by ten first, and then that one (1) ... ten tenths are one ... and then multiplied that (3) by ten and put it in three tenths.

This group of children, in spite of the fact that they came from four different year-groups and three different schools, all seemed to be at a similar transitional stage of development which seemed to be common among secondary school children.

Decimals as the result of a division

There seemed to be a marked reluctance to admit that the answer to the division of one whole number by another might be expressible as a number containing decimals, or even one containing fractions. This may well have been because pupils were still very much tied to the concrete idea of division as the sharing of objects such as sweets.

In the written test when asked to 'divide by 20 the number 24', (14d(i)) the responses were:

Table 4.2   Response percentages to Question 14d(i) by year

	12 yrs	13 yrs	14 yrs	15 yrs
there is no answer	15	12	11	6
1, or 1 remainder 4	19	12	10	4
1.4	8	16	14	15
$1\frac{4}{20}$ or $1\frac{1}{5}$	2	3	5	2
1.2	9	13	28	34

Dividing a smaller number by a larger one caused even more problems. When asked to 'divide by twenty the number 16', (14d(ii)) the answers were:

Table 4.3   Response percentages to Question 14d(ii) by year

	12 yrs	13 yrs	14 yrs	15 yrs
there is no answer	51	47	43	23
0, or 0 remainder 16	3	3	1	0
0.16	4	5	5	6
$\frac{16}{20}$ or $\frac{4}{5}$	1	0	2	1
0.8	7	12	25	36

Here there is a fairly clear indication that many children regard the division of a number by a number larger than itself as illegitimate, presumably in the sense that 16 sweets cannot fairly be shared among 20 people.

It does seem likely that mathematical experience in the secondary school helps in both these questions, as the rises in facility across the year-groups are higher than average.

In the different example below where 'opting out' was not encouraged, children found a way out by choosing to reverse the order of the numbers to give an answer greater than one.

18b(iii)   Ring the number you think is NEAREST IN SIZE to the answer (do *not* work out the sum)

$$59 \div 190 \rightarrow .003 \ / \ .03 \ / \ .3 \ / \ 3 \ / \ 30 \ / \ 300 \ / \ 3000$$

Table 4.4   Response percentages to Question 18b (iii) by year

Answer	12 yrs	13 yrs	14 yrs	15 yrs
.3	15	10	13	22
3	27	26	36	25

These answers were similar to those given in the interviews. For instance Kevin (14, 4th year), when asked to divide a smaller number by a larger one simply said 'You can't' while Cecilia (12, 2nd year), added '20 can't go into 16'. Kevin gave 1.4 as the answer for 24 divided by 20 since '20 goes into 24 once and there are four left over'.

Effect of multiplication and division by numbers less than one

It was clear that the idea that 'multiplication always makes it bigger, division always makes it smaller' was very firmly entrenched.

The question below illustrated this.

15.   Ring the one which gives the BIGGER answer:
   (a) 8×4 or $8 \div 4$
   (b) 8×0.4 or $8 \div 0.4$
   (c) 0.8×0.4 or $0.8 \div 0.4$

Table 4.5   Response percentages to Question 15 by year

	12 yrs	13 yrs	14 yrs	15 yrs
$\times, \div, \div$	13	8	15	18
$\times, \times, \times$	50	58	47	30

Fung Mei (13, 2nd year) was prepared to change her mind when asked to think again about her initial ringing of all the multiplications.

   I   Why did you choose the others first time?
   F   'Cos they're times and it always seems more than divide.
   I   Why isn't that true here?
   F   They've got decimals — in 8×0.4 you times by a little, in $8 \div 0.4$ you share between a little so each person gets more.

I   What if I asked you which was biggest, 0.8 × 1.2 or 0.8 ÷ 1.2?
F   That one (0.8 × 1.2)
I   It's got a decimal though.
F   'Cos it's got a whole number as well it makes it bigger.

The 'multiplication makes it bigger' syndrome also seems to affect the choice of the appropriate number operation.

19.   Ring the CALCULATION you would need to do to find the answer:
    C.   The price of minced beef is shown as 88.2 pence for each kilogram. What is the cost of a packet containing 0.58 kg of minced beef?

$$0.58 \div 88.2 \qquad 88.2 - 0.58 \qquad 0.58 - 88.2$$
$$0.58 \times 88.2 \qquad 88.2 + 0.58 \qquad 88.2 \div 0.58$$

    E.   My car can go 41.8 miles on each gallon of petrol on a motorway. How many miles can I expect to travel on 8.37 gallons?

$$8.37 \div 41.8 \qquad 41.8 - 8.37 \qquad 8.37 - 41.8$$
$$8.37 \times 41.8 \qquad 41.8 + 8.37 \qquad 41.8 \div 8.37$$

The costs given in C and E were realistic at the time the testing was carried out. In both these questions the correct response was multiplication, but the percentage choosing correctly was much lower in C where one of the numbers was less than one.

Table 4.6   Response percentages to Questions 19C and 19E by year

		12 yrs	13 yrs	14 yrs	15 yrs
19(C)	× (correct)	18	17	21	29
	÷ (incorrect)	37	39	48	42
19(E)	× (correct)	32	42	54	53
	÷ (incorrect)	29	20	22	18

In 19(C) children knew they were expecting a smaller answer than 88.2 and therefore tended to opt for a 'division' rather than a multiplication.

The infinite nature of the set of numbers expressible as decimals

The single question concerning this was:

12e   *How many* different numbers could you write down which lie between 0.41 and 0.42?

Table 4.7   Response percentages to Question 12e by year

	12 yrs	13 yrs	14 yrs	15 yrs
infinitely many, more than you can count	7	7	16	16
lots, hundreds	5	3	5	4
8, 9 or 10	22	39	36	38
1	17	8	8	9
0	9	5	4	2

These results were quite close to those which showed the ability to use the concept of 'infinitely many' in the context of graphs and fractions.

Some children like Tim (13, 3rd year, and surprisingly from a 'bottom band' class) had clearly met the idea of infinity before.

T 10, I expect . . . it could be thousands 'cos you can keep on doing it down even further – shall I put . . . er . . . in . . . *infernal,* is it?

I Inf . . .?

T Infinite (spells it 'infenate').

I Who told you about infinite?

T My mum, she teaches me a lot – my dad's good at maths, my mum is good at spelling and English . . . I can't remember where I learnt it . . . Dad does astronomy – infinite numbers of stars and things . . . he's a builder and decorator . . . it's not exactly his hobby (astronomy) but he's just interested in it.

The application of decimals

Children were asked to 'write a story' to see if they could think of circumstances in which they might use decimals.

20. One story which matches this sum is:

$6 + 2 = 8$ Linda had 6 records. She got another 2 for her birthday. So now she has 8.

Write a story which matches this sum:

$6.4 + 2.3 = 8.7$ *Story:*

Age	12	13	14	15
Facility	33	39	40	41 per cent

The answers were counted as 'correct' provided they represented quantities that could realistically be measured in this way (e.g. 6.4 gallons of petrol, kilogrammes of sugar, seconds etc.) but not if the units were taken as apples, sweets, books etc. and not even if they were used for money as in the expression £6.4 (although a conversion to £6.40 added to £2.30 was accepted).

The saddest response came from Raymond (12, 2nd year):

I (after long pause) Might you use decimals in a job?

R No, I don't think so.

I Or at home? (long pause) What about in any other lessons in school?

R Not really . . . you only do them in maths lessons.

Decimals – routine computation

All multiplication and division items involved multiples of powers of ten and hence could be done mentally by children who understood the process. Although the substitution of rules and algorithms for understanding worked on occasion, it was more than likely to prove an unreliable strategy in the long run. For example 26 out of 34 children who were asked during the interview not surprisingly knew the 'add a nought' rule for multiplying by ten and its extension to hundreds etc. However five of the 26 made mistakes in using it for whole numbers, and 15 proceeded to apply it inappropriately to decimals.

Only eight children out of the 39 children interviewed volunteered the 'move the figures (or the point) up' rule for multiplying decimals by ten, and four of the eight applied it wrongly in one or more question.

Twelve children attempted to either multiply or divide by powers of ten using long multiplication or division, but ten of these made errors in doing so. A typical error was made by Tim (13, 3rd year) who wrote

$$
\begin{array}{r}
327 \times \\
10 \\
\hline
327 \\
3270 \\
\hline
3597
\end{array}
$$

thus confusing the result of multiplication by zero with that by one. Shakeel (12, 2nd. year) could correctly multiply by ten by long multiplication, but had difficulty in placing the decimal point when working with hundreds, e.g.

$$
\begin{array}{r}
2.3 \\
100 \\
\hline
230 \\
000 \\
000 \\
\hline
23.000
\end{array}
$$

Results of the written tests: The levels

From the written test results, a set of items was identified which appeared to form a relatively homogeneous group, and these were divided into six levels by the procedure which is explained in Chapter 1.

A brief description of the levels is given in Fig. 4.1. The bands on the left show the spread of facility for items in each level for each of the ages 12, 13, 14, 15 years.

These facility bands overlap, due to the fact that the relative difficulty of the items sometimes changed from year to year.

Level 1: Place value in whole numbers up to thousands

Typical questions in this group were:
10b (i)   Ring the BIGGER number: 20 100 or 20 095

Age	12	13	14	15
Facility	86	89	91	94 per cent

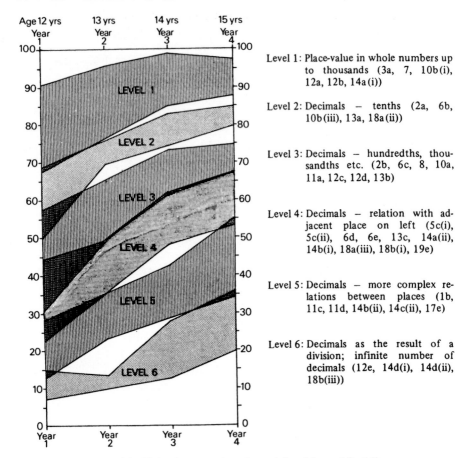

Age 12 yrs   13 yrs   14 yrs   15 yrs

Level 1: Place-value in whole numbers up to thousands (3a, 7, 10b(i), 12a, 12b, 14a(i))

Level 2: Decimals — tenths (2a, 6b, 10b(iii), 13a, 18a(ii))

Level 3: Decimals — hundredths, thousandths etc. (2b, 6c, 8, 10a, 11a, 12c, 12d, 13b)

Level 4: Decimals — relation with adjacent place on left (5c(i), 5c(ii), 6d, 6e, 13c, 14a(ii), 14b(i), 18a(iii), 18b(i), 19e)

Level 5: Decimals — more complex relations between places (1b, 11c, 11d, 14b(ii), 14c(ii), 17e)

Level 6: Decimals as the result of a division; infinite number of decimals (12e, 14d(i), 14d(ii), 18b(iii))

Fig. 4.1   Range of facilities for questions in each level (ages 12—15).

7.   This meter counts the people going into a football stand:

| 0 | 6 | 3 | 9 | 9 |

After *one* more person has gone in, it will read:

| | | | | |

Age	12	13	14	15
Facility	68	77	86	88 per cent

12b.   Write down any number between 4100 and 4200

Age	12	13	14	15
Facility	69	80	85	88 per cent

The questions involved fairly straightforward examples of the 'naming' of places and the type of simple relationships between places which are exemplified in the items quoted above.

Level 2: Decimals – tenths

Typical questions in this level were:

13a.

This is 1 square unit

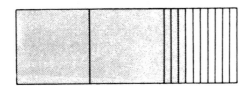

The area shaded is ⌐ . ⌐ square units. (Give your answer as a DECIMAL.)

Age	12	13	14	15
Facility	69	76	81	84 per cent

6b.

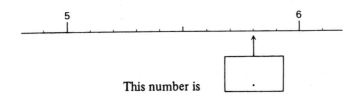

This number is ⌐ . ⌐

Age	12	13	14	15
Facility	62	74	83	85 per cent

10b (iii)   Ring the BIGGER number: 4.06 or 4.5

Age	12	13	14	15
Facility	66	72	83	80 per cent

2a.   0.2   The 2 stands for 2 . . . . . .

Age	12	13	14	15
Facility	63	70	73	79 per cent

These items were all concerned with either the straightforward naming of the tenths-place or with the meaning of this in a simple visual representation. Other items involving the relationship between tenths and units or between tenths and hundredths occur in harder levels (e.g. 5c (ii) quoted in level 4 and 11d quoted in level 5 below).

Level 3: Decimals – hundredths, thousandths, etc.

Typical items include:

11a. Six tenths as a decimal is 0.6. How would you write as a decimal three hundredths?

Age	12	13	14	15
Facility	50	60	59	57 per cent

12d. Write down any number between 0.41 and 0.42.

Age	12	13	14	15
Facility	37	49	66	71 per cent

6c.

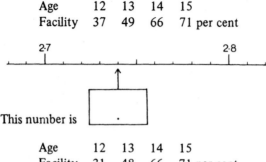

Age	12	13	14	15
Facility	31	48	66	71 per cent

As before it turned out that the 'scale-reading' item, 6c., showed a much greater increase over the four years than did most other items, perhaps because of increased experience of reading scales in science subjects. However other 'interpolation' items which required the idea of the 'hundredths' representing divisions between two adjacent 'tenths' had a similar pattern. The straightforward item 11a. which required naming, together with the use of zero as a place-holder, was found comparatively very much easier by first and second year pupils, who may have learnt this by rote, but surprisingly more difficult than the other two items by third and fourth year children.

Level 4: Decimals – relation with adjacent place on left

Typical examples include:

18a (iii) Ring the number NEAREST IN SIZE to
0.18 → 0.1 / 10 / 0.2 / 20 / .01 / 2

Age	12	13	14	15
Facility	44	48	61	59 per cent

14a (ii)   Multiply by ten: 5.13 → . . . . . .

Age	12	13	14	15
Facility	37	42	58	65 per cent

5c (ii)   Add one tenth: 2.9 → . . . . . .

Age	12	13	14	15
Facility	38	44	51	59 per cent

6d.

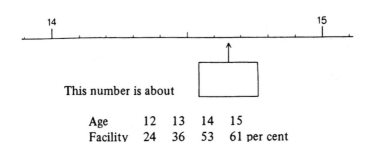

This number is about

Age	12	13	14	15
Facility	24	36	53	61 per cent

Most items in this group required the idea that ten in one place would be equivalent to one in the next place on the left which is the basis of the 'carrying' process in the addition algorithms.

Again the 'scale-reading' items were found relatively much more difficult than the others by first and second years.

The second example, 14a (ii), could have been done purely by rote with little understanding. However the fact that the item correlates well with, and shows a similar pattern to, other items in this level supports the evidence gained from the interviews that on the whole children with little understanding did not manage to arrive at a correct answer even by rote.

The example 6d. is more difficult than 6c. in the previous level as it requires not simply counting divisions, but the knowledge that there would be ten hundredths between 14.6 and 14.7 and hence the point half way along would be 14.65.

Level 5: Decimals — more complex relations between places

Typical examples include:

11d. *Four tenths* is the same as . . . . . . hundredths

Age	12	13	14	15
Facility	28	31	42	40 per cent

14b (ii)   Divide by one hundred: 3.7 → . . . . . .

Age	12	13	14	15
Facility	24	27	34	41 per cent

1b.   5214   The 2 stands for 2 *HUNDREDS*

521 400   The 2 stands for 2 . . . . . .

Age	12	13	14	15
Facility	22	32	31	43 per cent

This group of questions was rather diverse, but it included relations between two non-adjacent places (as in 14b(ii) above), as well as the ability to reverse the reasoning in level 4 questions to arrive at the fact that one in a given place is equivalent to ten in the next place along on the right (as in 11 d). This latter process is that of 'borrowing' required in the subtraction algorithm.

The item 1b. seems at first rather out of place here, but essentially a not too dissimilar step is involved in going from the idea of '20 thousand' (i.e. 20 in the 'thousands' place) to '2 × ten thousand'.

Item 11d. embodies the concept of 'equivalent fractions', which also appears in the fractions paper and seems to be of a similar order of difficulty there (see the next chapter).

Level 6: Decimals as the result of a division; infinite number of decimals

The items in this level were:

14d.   Divide by 20

(i)   24 → . . . . . .

(ii)   16 → . . . . . .

Write 'NO' if you think there is no answer.

Age	12	13	14	15
Facility (i)	9	13	28	34 per cent
Facility (ii)	7	11	25	36 per cent

18c (iii)   Ring the number you think is NEAREST IN SIZE to answer (do NOT work out the sum): 59 ÷ 190 → .003 / .03 / .3 / 3 / 30 / 300 / 3000

Age	12	13	14	15
Facility	15	10	13	22 per cent

12e.   How many different numbers could you write down which lie between 0.41 and 0.42?

Age	12	13	14	15
Facility	12	10	21	20 per cent

(The percentage 'correct' for item 12e. includes children who gave an answer like 'lots' or 'hundreds'.)

Results of the written tests − The children

A child was said to have attained a level if he or she answered two-thirds of the items correctly at that level. Fig. 4.2 shows the proportions of children in each year group who were judged to have attained each level. Ninety-three per cent of the pupils were 'scale-types' (i.e. attained a level only when all easier levels had been attained); the remaining seven per cent are shown on the diagram at the highest level which they attained, even though one or more previous ones had not been attained.

Implications for teaching

The major point to notice in Fig. 4.2 is that all seven levels of understanding are found in each of the years 1, 2, 3 and 4 of the secondary school, although the proportions differ from year to year.

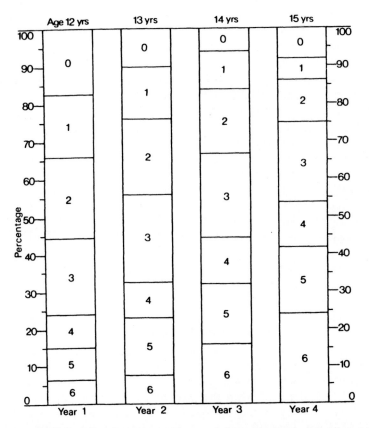

Fig. 4.2   Percentage of children in each year of the secondary school (aged 12−15) who are judged to have attained each level. (The percentage labelled 0 did not attain any level.)

Thus in a truly mixed ability first year class of 25 children, about four children would be expected to have little understanding of place value even in whole numbers (i.e. Level 0), whereas the number would have decreased to two in a similar fourth year class. In the same way about two children in the first year class, and six in the fourth year class, would be expected to understand thoroughly the fundamental ideas of the decimal system.

One could generalise by concluding that the top 50 per cent of pupils are likely by the time they leave school to have a reasonable if not complete understanding of decimals, whereas the lower 50 per cent of pupils still have considerable gaps in their understanding. However this does not necessarily mean that they would not be able to cope in a more concrete situation where the decimals referred to measures of money, length etc.

Since all these stages are present in any year, it is not appropriate to recommend a programme of work for first year children followed by another for second years and so on.

The major need is for careful diagnosis of what progress each individual has made using the sort of questions included here together with a corresponding set of units of work.

It is above all clear that the learning of whole numbers and decimals is not just a matter of recalling some place-names and a few rules of computation, as it often appears to be from the textbooks. Indeed the children who did rely blindly on rules more often misapplied them than not.

Instead, it involves internalising a whole chain of relationships and connections, some within the place-structure itself (e.g. 0.9 is equivalent to 0.90), some linking to other concepts like those of fractions (e.g. the notion of one hundredth and its relationship to one tenth), some visual correspondences and some connecting to applications in the 'real' world.

One particular area where children seemed weaker than expected was with whole numbers over a thousand. It may be that secondary school teachers tend to assume children have mastered these ideas before the age of 11, but this seems not to be generally the case.

It was clear that many children still needed visual models of tenths, hundredths and so on, to bring out the relationships in a more concrete way. The use of Dienes Base 10 blocks (with a thousand block as unit) or even just squared paper divided, or actually cut, into units, rows of ten, squares of a hundred and so on would obviously help.

The advent of the calculator will obviously make children more familiar with decimal representation, but without careful structuring of the work it may just be used to produce meaningless answers which can be copied down faithfully to eight decimal places. However if the right questions are asked of children (e.g. Why does $2 \div 8$ give 0.25 when the child may have expected the answer to be 4? Can he explain by cutting up squared paper if necessary? Is 3.9999999 the same as 4? If not, how different is it?) then obviously the calculator will prove a most useful teaching aid.

'Pencil and paper' computation techniques with complicated whole numbers or decimals are no longer essential, but anyone who understands enough about decimals to be able to use a calculator sensibly should be able, given time, to work out most of the 'rules' from first principles anyway, as a number of children managed to do during the interviews.

It is certainly to be hoped that the presence of calculators will shift the emphasis from routine techniques, which did not seem to be performed very reliably, to the understanding of the principles, especially since this latter aspect seems to have been rather neglected at secondary school level in this particular area.

5   Fractions

Fractions (naming and operations on them) formed the basis of two tests given by the CSMS Mathematics team, one for 12 and 13 year olds, the other for 14 and 15 year olds. The topic of fractions is a part of that mathematics which needs its own specific language and it was thought that the language which might have been part of the common experience of third and fourth year pupils would not necessarily have been taught to first and second years.

The questions designed for the younger children could all be successfully completed by addition and subtraction. The paper for the older children contained some items in common with this paper but also presented problems which required multiplication and division, e.g. questions about the area of a rectangle. Each test paper presented problems (P notation below) and in addition a set of computations (C notation below), designed to mirror the problems, was also given to the children. Thus a problem might be:

P22.   Shade in $\frac{1}{3}$ of the dotted section of the disc. What fraction of the whole disc have you shaded?

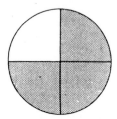

The comparable computation simply reads: $\frac{1}{3}$ of $\frac{3}{4}$ = . . . (C10). The computations were given to the children immediately after they had finished the set of problems and though it was thought that the children would find the computations easier than the problems, this proved not always to be so.

Topics covered

12–13 year olds

The problem paper involved questions posed either in words or diagramatically. Diagrams were used for the naming of parts, recognition of one-half and as in (P22) above. A fractional name can be given to a subset of a collection of objects as well as to a part of a whole (e.g. a pie or rectangle) and items of both sorts were included.

A major part of the early work on fractions carried out in schools concerns the idea of equivalence, e.g. $\frac{1}{2}$, $\frac{2}{4}$, $\frac{3}{6}$ ... Questions involving equivalents of well-known fractions such as $\frac{1}{2}$ and also less familiar ones such as $\frac{2}{7}$ occurred in many of the problems. Addition of fractions with the same denominator and where one denominator was a multiple of the other also appeared. Two questions tested the ability to subtract a fraction from 1. Opportunities for using multiplication and division were provided but it was realised that other methods were available to the children.

14 and 15 year olds

The examples for the older children included some of those described above and also some attempt to test how well the children could generalise the concept of 'fraction'. For example, the older children were asked 'How many fractions lie between $\frac{1}{4}$ and $\frac{1}{2}$?' Multiplication was set in the context of finding area and also in rate. The latter for example was applied to the conversion of miles to kilometres. The children were not asked to divide a mixed fraction by another but were asked to choose between four alternative ways of doing $3\frac{1}{3} \div 2\frac{1}{3}$ (page 72).

Results

Comparison between computation questions and problems

Not all of the computations appeared on the two test papers but certain addition and subtraction computations were common to both. In each of these the percentage of first year pupils succeeding was always higher than in every other year. Even when the computation involved mixed numbers as in $32\frac{2}{3} + 5\frac{1}{4}$ for example, the percentage of first year children successfully completing this item was twice that in any other year. The youngest children had been taught the addition and subtraction rules more recently and therefore did better on these items. In many cases the problems proved easier than their comparable computations, pointing to the fact that the children used strategies other than the algorithms they had been taught. There appeared to be no connection in many children's minds between the problem and the 'sum' since they could successfully deal with the problem but could not apply the same method to the computation. It was as if two completely different types of mathematics were involved, one where the children could use common-sense, the other where they had to remember a rule.

Fractions and whole numbers

The child feels relatively secure when working within the set of whole numbers and when bound by the restrictions imposed by them. The fact that some of these restrictions do not apply within the set of fractions and indeed that fractions are invented in order to extend the number system beyond that which is necessary for counting, often escapes him. For example asked for an answer to 3 ÷ 5 many interpreted it as 5 ÷ 3, presumably because they were applying '5 into 3 won't go'. In the computation form (3 ÷ 5) the answers given were as follows:

Table 5.1   Different responses to 3 ÷ 5 (percentage)

	$\frac{3}{5}$ or .6	$1\frac{2}{3}$	1 rem 2	$\frac{5}{3}$ or $1\frac{2}{3}$
12 yrs	35	5.3	18.3	3.3
13 yrs	31	9.4	17.5	8.7

Between 25 and 30 per cent of each year divided the smaller number into the larger in some way.

Rules such as 'multiplying always makes it bigger' were applied by the children to fractions as in the following question given to the older group.

P23.

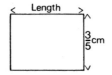

Area = $\frac{1}{3}$ square centimetre

Length = . . .

This was very difficult for the children. Many had forgotten how to find an area but others said '$\frac{3}{5}$ is bigger than the area, it cannot be done'. Only 7 per cent of the 14 year olds and 5.6 per cent of the 15 year olds successfully completed this item, but the vast majority did not attempt it.

Another sign of lack of ease with fractions is the insistence on giving an answer in *remainder* form rather than *fractional* form. In any sharing process the setting of the problem dictates whether a fractional answer or a remainder form is preferable. For example in a problem asking how many 4 cm bicycle spokes can be made from 17 cm of metal, $4\frac{1}{4}$ bicycle spokes is less sensible than 4 bicycle spokes and 1 cm of metal left over. In the following question however the fractional answer is more sensible:

P5.   A piece of ribbon 17 cm long has to be cut into 4 equal pieces. Tick the answer you think is most accurate for the length of each piece.

(a)   4 cm remainder 1 piece

(b)   4 cm remainder 1 cm

(c) $4\frac{1}{4}$ cm

(d) $\frac{4}{17}$ cm

Note that the child simply had to choose an answer. Many preferred the remainder version in (b).

Table 5.2   Responses to P5 (percentage)

	$4\frac{1}{4}$ cm	4 cm rem. 1 cm	4 cm rem. 1 piece	$\frac{4}{17}$ cm
12 yrs	43.0	37.4	8.5	9.3
13 yrs	54.4	29.8	6.8	6.5
14 yrs	61.4	26.6	4.2	5.8
15 yrs	61.4	27.4	3.7	6.0

A fraction of course involves two whole numbers which have to be dealt with as if they were irrevocably linked. When adding one looks at the denominators of the two fractions involved but only to guide one in changing the pair (numerator, denominator) into a more convenient form. A common mistake, on the test paper, when $\frac{2}{7}$ was given and two equivalents had to be found, was to regard the numerators as forming one number pattern and the denominators another but ignoring the ratio numerator:denominator. So in $\frac{2}{7} = \frac{\square}{14} = \frac{10}{\triangle}$, 21 and 28 were often given as replacements for $\triangle$. About 16 per cent of each year group gave either 21 or 28 as the answer.

Methods used by children

Labelling Parts

P20.   What fraction is shaded?

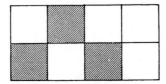

Questions of this type were almost invariably solved by counting the number of squares in the entire figure, counting those shaded and putting one number over the other. On interview, when asked for the fraction not shaded, few children subtracted $\frac{3}{8}$ from 1; they usually counted once again. A *half* seems to be very much easier to deal with than any other fraction, so that

is labelled after counting but

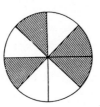

is almost immediately seen as $\frac{1}{2}$ and not four out of eight; this is possibly obtained by matching four to four.

There was some difficulty with the question (P22) 'Shade $\frac{1}{6}$ of the dotted section of the disc. What fraction of the whole disc have you shaded?'

Although many children could shade the correct area, when they counted the total available parts they included 'a' as equal to each of the other pieces ($\frac{1}{6}$ of $\frac{3}{4}$) and so obtained one out of seven pieces. Others, having been given $\frac{1}{6}$ as the fraction wanted, gave this as the answer, equating the fraction of the dotted section with the fraction of the whole disc. This tendency to ignore the question 'a fraction of what?' also occurred on what was the hardest question on the paper (discussed later) when the children stated $\frac{1}{2}$ was always bigger than $\frac{1}{4}$ although the fractions were of different amounts.

Use of diagrams

Diagrams often helped towards the solution of the problem. Thus, in the example cited previously, the success rate for the two forms of the question differed considerably.

P22.

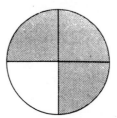

Shade in $\frac{1}{6}$ of dotted section. What fraction of the whole disc have you shaded?
Computation: $\frac{1}{6}$ of $\frac{3}{4}$

Table 5.3   Correct responses to P22 and computation (percentage)

	Problem	Computation
12 yrs	54.1	22.8
13 yrs	51.8	23.0
14 yrs	56.5	25.6
15 yrs	56.7	30.2

It was sometimes apparent on interview that the child needed a diagram to help him *see* what a word problem required. The following word problem appeared on both test papers:

P4.   5 eggs in a box of 12 are found to be cracked. What fraction of the box of eggs is cracked? What fraction of the box of eggs is not cracked?

Some children said they could not do this but when they had read the next problem which provided a rectangle divided into eight squares they extended the rectangle to count 7 and 12.

It was sometimes obvious that the child used a diagram to reinforce what he already knew. For example, faced with $\frac{2}{7} = \frac{?}{14}$ a child drew

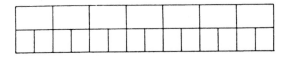

He already knew that each seventh provided two fourteenths otherwise he could not draw the diagram. However some children thought the diagram only 'worked' if each part on the top was replaced by two on the bottom so that they could not cope with turning tenths into thirtieths. As one said 'I don't have enough, I only have twenty'.

Equivalent fractions

Items which asked children to find equivalent fractions varied in difficulty, questions involving $\frac{1}{2}$ being the easier ones. Faced with $\frac{1}{3} = \frac{2}{?}$ and $\frac{5}{10} = \frac{?}{30}$ many children used a multiplying strategy on the first but chose to say '$\frac{5}{10}$ is $\frac{1}{2}$, so I want a half of 30 for the top of the second one'. Children did not necessarily multiply numerator and denominator by a number and so their difficulties resulted from their method and not their knowledge of multiplication facts. Some children, faced with $\frac{2}{7} = \frac{\square}{14}$, did not multiply by two but said '7 and another 7 make 14, so its 2 and another 2'. This method becomes cumbersome when the question is $\frac{2}{7} = \frac{10}{\triangle}$ since the number of sevens to be added together has to be remembered. Another error resulted from a desire to find a number pattern, e.g. '$\frac{2}{7} = \frac{10}{15}$ because 2 is 5 less than 7 and we want a number 5 less than 15'. This is again indicative of the view that a fraction is simply two whole numbers which can be treated separately.

The transitivity of the equal sign often escapes some children. Thus in $\frac{2}{7} = \frac{\square}{14} = \frac{10}{\triangle}$ the finding of the value for $\triangle$ was much more difficult than finding the value of $\square$. The child often tried to work with $\frac{4}{14} = \frac{10}{\triangle}$ and no whole number multiple of 4 gives 10. The inability to use $\frac{2}{7} = \frac{10}{\triangle}$ may arise because they view equality as relevant only to the two expressions on either side of the equal sign.

Equivalence occurred in the solution to the problem:

P24.   A relay race is run in stages of $\frac{1}{8}$ km each. Each runner runs one stage. How many runners would be required to run a total distance of $\frac{3}{4}$ km?

This was very much easier than the computation $\frac{3}{4} \div \frac{1}{8}$ and it became clear that the children were not using the operation of division to solve the problem. When interviewed many simply translated $\frac{3}{4}$ into $\frac{6}{8}$ and said 'six runners'. One could of course say 'two runners run $\frac{1}{8} + \frac{1}{8} = \frac{2}{8} = \frac{1}{4}$, so I need three pairs of runners', but the first equivalence strategy is much neater.

The hardest question on the problem paper for each group of children was the following item which required the child to apply conditions once he had ascertained the possibility that Mary has more money than John:

P17.   Mary and John both have pocket money. Mary spends $\frac{1}{4}$ of hers and John spends $\frac{1}{2}$ of his. Is it possible for Mary to have spent more than John? Why do you think this?

Table 5.4   Responses to P17 (percentage)

	12 yrs	13 yrs	14 yrs	15 yrs
Answer with Mary having twice more	1.6	2.3	1.9	3.7
Mary having more than John	32.1	38.8	36.7	46.0
Answer No, $\frac{1}{2}$ is bigger than $\frac{1}{4}$	41.5	34.3	27.6	19.1

Notice the number of children who realise that Mary may have had more to start with but cannot provide limits for the amount. Perhaps more important is the large number of children who feel that '$\frac{1}{2}$ is bigger than $\frac{1}{4}$' is a valid reason for rejecting the idea of Mary spending more than John.

A common error when dealing with equivalent fractions was to look at the size of the number in the numerator and the size of the one in the denominator and not at the ratio of the two. Twenty per cent of each year aged 12 and 13 denied the equality of $\frac{5}{20}$ and $\frac{1}{4}$, and 20 per cent likewise said that $\frac{4}{8}$ was larger than $\frac{2}{4}$.

Multiplication of fractions

The hardest group of problems for the 14 and 15 year olds did not appear on the test paper for the younger children. These involved multiplication and division of fractions, and over 40 per cent of each year omitted the items. Multiplication and division of mixed numbers gave the greatest difficulty as the children attempted to deal with the whole number parts separately. Thus when asked to compute $3\frac{3}{4} \times 2\frac{1}{2}$, 17 per cent of the 14 year olds and 16 per cent of the 15 year olds gave the answer $6\frac{3}{8}$.

A similar type of response was apparent in P29 where the child had to choose the appropriate method.

P29. $3\frac{1}{3} \div 2\frac{1}{5}$ is equal to:

(a) $(3 \div 2) + (\frac{1}{3} \div \frac{1}{5})$ (b) $(3\frac{1}{3} \div 2) + (3\frac{1}{3} \div \frac{1}{5})$

(c) $(3\frac{1}{3} \div 2) + \frac{1}{5}$ (d) $(3 \div 2\frac{1}{5}) + (\frac{1}{3} \div 2\frac{1}{5})$

Tick which one you think is correct.

A number of children (35 per cent of each year) opted for $(3 \div 2) + (\frac{1}{3} \div \frac{1}{5})$, the correct choice being made by only 10.7 per cent of the 14 year olds and 5.1 per cent of the 15 year olds. It is recognised that if they had been asked to do the computation, the children may have used an algorithm which gave the correct answer but the 35 per cent who opted for (a) certainly did not check the correspondence between this version and the rule.

Generalisation of fractions

The older children were given questions to test their ability to generalise the laws of fractions. Some of these items involved letters instead of numbers and in these cases some 60 per cent of the sample.omitted them. Although in the same test paper the success rate on comparisons was high, the lettered version had very little success.

Table 5.5   Generalised comparisons (percentage)

	$\frac{3}{4} > \frac{3}{5}$	$\frac{5}{7} > \frac{3}{7}$	$\frac{a}{d} > \frac{a}{b}$ when $b > d$
14 yrs	74.0	90.6	13.6
15 yrs	78.6	90.7	16.3

The children were asked how many fractions came between $\frac{1}{4}$ and $\frac{1}{2}$ and 55 per cent of the 14 year olds and 52 per cent of the 15 year olds gave a number less than twenty. Thirty per cent gave the answer 'one' which presumably was the fraction $\frac{1}{3}$ if they considered only fractions with numerator 1 and denominator as part of a number pattern. The children were asked to write down a fraction that came between $\frac{1}{2}$ and $\frac{2}{3}$; the replies are shown in Table 5.6.

Table 5.6   A fraction between 1/2 and 2/3 (percentage)

	Correct (any)	Correct (7/12)	1/3	1/4	3/4
14 yrs	26	13	27	10	10
15 yrs	21	12	24	10	10

It is highly likely that many children did not realise that a search for equivalent fractions would provide a solution to the problem. The answer $\frac{1}{3}$ highlights again the methods by which comparisons are made. Comparing the numerator would give $\frac{2}{3} > \frac{1}{3}$ and comparing the denominator would give $\frac{1}{3} > \frac{1}{2}$; the child is not aware of the inconsistency.

The child is often very restricted in the way he sees problems involving fractions, e.g. he does not say 'what does this mean?' but 'what do I do when that sign (e.g. ÷) appears?'. For example, faced with this problem which does not require computation but careful thought on what would happen if the operation was carried out, the success rate was low:

P25. $\frac{1}{17} \div \frac{2}{5} = \frac{5}{34}$ Then $\frac{1}{17} \div \frac{4}{5}$ is:

(a) Twice $\frac{5}{34}$ (b) $\frac{4}{5}$ of $\frac{5}{34}$ (c) $\frac{5}{4}$ of $\frac{5}{34}$ (d) Half of $\frac{5}{34}$

Tick which one you think is correct.

The results are shown in Table 5.7

Similarly, since fractions are often introduced by using a 'whole' and parts of identical shape and size, when a problem is presented using parts of two different

Table 5.7   Choices on P25 (percentage)

	(a)	(b)	(c)	(d)
14 yrs	24.4	11.4	11.4	21.1
15 yrs	28.8	14.0	6.5	19.5

sizes (albeit one being twice as large as the other) the child is unable to use the smaller as a 'part' and is distracted by the presence of another alternative. Thus in the following question using a square rather than a triangle as the unit for counting parts was preferred by many:

P20.   I am putting tiles on the floor; they are shown shaded. What fraction of the floor has been tiled?

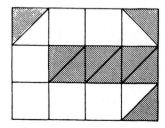

Table 5.8   Replies to P20 (percentage)

	$\frac{9}{24}$ or $\frac{3}{8}$	$4\frac{1}{2}/12$
12 yrs	30.5	18.7
13 yrs	28.8	22.0
14 yrs	37.0	21.4
15 yrs	42.3	20.5

The same lack of flexibility is demonstrated in the following four questions, two problems and two computations.

(a)   How many bicycle spokes $10\frac{1}{2}$ cm long can be cut from a piece of wire 40 cm long?

(b)   What length of wire is left over?

(c) $3 \times 10\frac{1}{2}$

(d) $40 \div 10\frac{1}{2}$ (3 rem. $8\frac{1}{2}$ counted as correct)

	12 yrs	13 yrs	14 yrs	15 yrs
(a)	69.5	74.1	77.6	79.1
(b)	52.8	56.3	64.6	65.2
(c)	76.0	81.6	79.9	84.2
(d)	25.1	20.1	20.8	27.4

Since whatever method was used to solve the problems could have been applied to the second computation it would seem likely that the children did not recognise it as essentially the same question. The two computations were adjacent on the paper and again the children did not connect the two operations. Some children may have rejected 3 rem 8½ as an inadequate answer but only 8−16 per cent gave the answer $3\frac{17}{21}$. The division algorithm is very difficult to apply (30 per cent of the sample could deal with $\frac{3}{4} \div \frac{1}{8}$) and probably any computation which seemed to require its use was likely to upset the child.

Errors

A very common error in the addition of fractions was to use a rule 'add tops add bottoms'. This occurred on each computation involving the addition of two fractions and was more prevalent on examples where the two denominators were different, e.g.

Table 5.9   Incorrect addition − computation (percentage)

	12 yrs	13 yrs	14 yrs	15 yrs
$\frac{3}{8} + \frac{2}{8} = \frac{5}{16}$	8.5	19.7	14.3	16.7
$\frac{1}{3} + \frac{1}{4} = \frac{2}{7}$	18.3	29.1	21.8	19.9

It is also interesting to note that this particular error occurred more when the question was posed in computation form than in problem form e.g. P19.

P19.   In a baker's shop $\frac{3}{8}$ of the flour is used for bread and $\frac{2}{8}$ of the flour is used for cakes. What fraction of the flour has been used?

Table 5.10   Responses to P19 and comparable computation (percentage)

	12 yrs	13 yrs	14 yrs	15 yrs
$\frac{5}{16}$ problem	4.9	7.4	6.8	6.0
$\frac{5}{8}$ problem	77.0	73.0	80.8	77.2
$\frac{5}{8}$ computation	77.6	66.3	71.8	67.9

Groups and levels of understanding

The method of analysis used on the data in order to obtain groups of items has been explained in Chapter one. Four groups of problems (the computations were dealt with separately) were formed from the data available from each test paper. The facility range for each is shown in Table 5.11. Note that those problems which appeared on both the paper for the younger children and on that for the 14−15 yr old sample were successfully completed by rather more of the older group.

Comparison of performance between the two age groups

The test paper given to the older children had fewer questions requiring them simply to name a fraction, and there is therefore no level of understanding for this

Table 5.11   Levels for fraction problems

Test 1 (12–13 year olds)			Test 2 (14–15 year olds)		
Level	Pass mark	Facility Range (percentage)	Level	Pass mark	Facility Range (percentage)
1	$\frac{2}{3}$	87–93	1	$\frac{7}{10}$	72–85
2	$\frac{7}{10}$	67–80	2	$\frac{5}{8}$	52–66
3	$\frac{7}{11}$	46–67	3	$\frac{2}{4}$	31–40
4	$\frac{3}{4}$	24–38	4	$\frac{3}{5}$	6–15

Note. Error: 2 per cent

Table 5.12   Levels for fraction computation

Test 1 (12–13 year olds)			Test 2 (14–15 year olds)		
Level	Pass mark	Facility Range (percentage)	Level	Pass mark	Facility Range (percentage)
1	$\frac{3}{5}$	60–80	1	$\frac{3}{5}$	64–82
2	$\frac{3}{5}$	45–55	2	$\frac{3}{5}$	38–47
3	$\frac{3}{5}$	32–38	3	$\frac{2}{4}$	23–32
4	$\frac{2}{2}$	8–24	4	$\frac{3}{5}$	11–17

group which is comparable to Level 1 for the younger children. There were no
questions which required division and multiplication of fractions on the paper given
to the younger children so consequently Level 4 for the older children has no
equivalent in the levels outlined for the younger children. Many problems were
common to both papers, and the older children performed slightly better on each
one. This is in direct contrast to the computation items involving addition and
subtraction when the youngest children performed best. In Fig. 5.2 on page 80
1st and 2nd years can be compared with each other but not with the 3rd and 4th
years since they have different levels, defined in different ways.

Implications for teaching

The notation of fractions is introduced in most British primary schools and the
child is expected to know the meaning of a half and a quarter at a very early age
even if he does not know how to add and subtract fractions. Children in the
secondary school appear to recognise equivalents of a half and to deal with them in
a different way to other fractions (although 7 per cent of our 13-year-old sample
did not know that it meant one of two equal pieces). Many children do not feel
confident in the use of fractions and try whenever possible to apply the rules of
whole numbers to operations on fractions. They much prefer a remainder type
answer than one which states a fraction and generally seem unaware that working
within the set of fractions enables them to manipulate numbers in a far less
restricted way than when they only had whole numbers with which to work. They
still appear to be fixed within rules which apply to whole numbers, e.g. division of a

Table 5.13   Problem levels (12—13 year olds)

Level 0	Unable to make a coherent attempt at level 1 problems	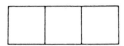 Shade in two-thirds
Level 1	The meaning of a fraction using pieces, $\frac{1}{2}, \frac{1}{5}, \frac{2}{3}$	
Level 2	The meaning of a fraction as a subset of a set or naming given configuration of pieces. Equivalent fractions obtained by doubling.	John wins 1/3 of these marbles. Draw a ring round his marbles.
	Addition of two fractions with same denominator	$\frac{1}{3} = \frac{2}{?}$
Level 3	Using equivalence to name parts, with familiar fractions or when diagram provided. Equivalent fractions not obtained by doubling or less familiar, e.g. $\frac{2}{7} = \frac{\square}{14}$	Shade in two-thirds $\frac{4}{12} = \frac{1}{?}$
		Shade in 1/6 of the dotted section of the disc. What fraction of the whole disc have you shaded?
Level 4	Questions where more than one operation is needed, e.g. an equivalence followed by addition or subtraction	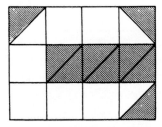 I am putting tiles on the floor. They are shown shaded. What fraction of the floor has been shaded?

Table 5.14   Problem levels (14–15 year olds)

Level 0	Unable to make a coherent attempt at level 1 problems	
Level 1	The meaning of a fraction, seen as part of a whole, no equivalence needed. Equivalent fractions obtained by doubling. Addition of two fractions with the same denominator	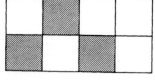 What fraction is shaded? $\frac{1}{3} = \frac{2}{?}$
Level 2	Equivalent fractions not obtained by doubling. Using equivalence to name parts, with familiar fractions or when diagram provided	 $\frac{2}{3} = \frac{?}{15}$
Level 3	Questions where more than one operation is required e.g. equivalence followed by addition or subtraction	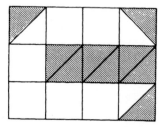 I am putting tiles on the floor. They are shown shaded. What fraction of the floor has been shaded?
Level 4	Division and multiplication of fractions	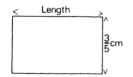 Area $= \frac{1}{3}$ square centimetre Length $= \ldots$

small number by a larger is impossible and multiplication makes bigger. The conventions of fractional notation are usually introduced with reference to part of a whole or the number in a subset in relation to the number in the entire collection of objects. The conventions are presented in two ways:

(1)   given the fraction show the concrete referent;
(2)   shown the parts of a whole or subset of a collection, label it with the correct fractional name.

The use of fraction words or symbols in relation to concrete referents appears in three different levels in the hierarchy described in this chapter. To name a part when the whole object is divided into exactly the same amount of parts as the denominator of the fraction being sought is relatively easy. When however the

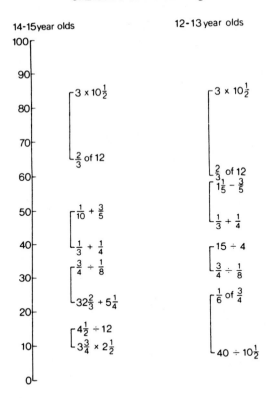

14-15 year olds

100

90

80 ⌐3 × $10\frac{1}{2}$

70

60 ⌐$\frac{2}{3}$ of 12

50 ⌐$\frac{1}{10} + \frac{3}{5}$

40 ⌐$\frac{1}{3} + \frac{1}{4}$

30 ⌐$\frac{3}{4} ÷ \frac{1}{8}$

20 ⌐$32\frac{2}{3} + 5\frac{1}{4}$

10 ⌐$4\frac{1}{2} ÷ 12$
    ⌐$3\frac{3}{4} × 2\frac{1}{2}$

0

12-13 year olds

⌐3 × $10\frac{1}{2}$

⌐$\frac{2}{3}$ of 12
⌐$1\frac{1}{5} - \frac{3}{5}$

⌐$\frac{1}{3} + \frac{1}{4}$

⌐15 ÷ 4

⌐$\frac{3}{4} ÷ \frac{1}{8}$

⌐$\frac{1}{6}$ of $\frac{3}{4}$

⌐40 ÷ $10\frac{1}{2}$

Fig. 5.1   Computation levels.

problem involves finding the number which is one third of twelve, 20 per cent of the younger children cannot do it. When the problem is to find two-thirds and the whole is divided into six parts, 36 per cent of the younger children fail. There is obviously a considerable increase in complexity between naming when no equivalence is needed and when it is. Equivalent fractions form the basis on which the operations of addition and subtraction are built but except for those which require doubling they appear to be understood by less than 60 per cent of the first and second years. Many children seem to view the fraction as two unrelated whole numbers and deal with each separately as in the example $\frac{2}{7} = \frac{\square}{14} = \frac{10}{\triangle}$.

There is also evidence that the child concentrates on the two numbers in the ratio and judges the size of the entire fraction on the *comparative* size of the denominators or numerators, e.g. 20 per cent of the 12−13 year olds stated $\frac{4}{8}$ was larger than $\frac{2}{4}$.

The ability to solve addition and subtraction computations *declines* as the child gets older. The ability to solve the problems does *not decrease* with age and one is left with the hypothesis that the problems are solved without recourse to the computational algorithms. Many children do not in fact seem to connect the algorithms with the problem solving and use their own methods. Most of the computation rules are only short cuts for what the child is using, e.g. 5 times 2 is a

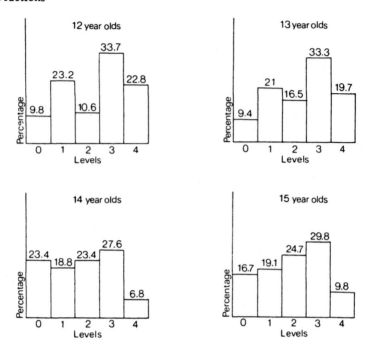

Fig. 5.2   Percentage of children at each level in the 'problem' hierarchy.

short cut for 'a two and another two and another two etc.' The child does not rewrite the problem in the form of a 'sum' but uses methods which appear to fit the situation.

Diagrams appear to be useful; sometimes they are used directly but on other occasions the child appears to need one to check that his answer is feasible. If he is required to provide a fraction name, the provision of squares which can be counted appears to be an advantage.

The hardest level in the hierarchy for older children is composed of multiplication and division problems, these are successfully completed by few, yet the computation $10\frac{1}{2} \times 3$ is the easiest on the test paper. One suspects that it is done by repeated addition since one of the elements is a whole number. If the question is dealt with in two parts i.e. $(3 \times 10) + (3 \times \frac{1}{2})$, the answer is correct. Contrast this with $3\frac{1}{2} \times 2\frac{1}{2}$ which cannot be translated into $(3 \times 2) + (\frac{1}{2} \times \frac{1}{2})$. The meaning of multiplication is firmly rooted in the child's experience of whole numbers where the operation can always be replaced by repeated addition. If the child sees 4×3 as four groups of three objects (which can be spaced out and counted) the meaning he attaches to $\frac{1}{3} \times \frac{5}{6}$ is unclear (one third part of six pieces of a pie?). Certainly he can no longer count to obtain an answer and if in fact he has avoided the operation of multiplication altogether by using repeated addition, there is no way he can deal with $\frac{1}{3} \times \frac{5}{6}$. Multiplication of fractions cannot be dealt with by the use of naive and intuitive methods and is therefore based very much more on 'rule learning' than some other aspects of mathematics.

Concrete embodiments used when multiplication of fractions is introduced are usually restricted to the area of rectangles when the dimensions are no more complicated than wholes and halves because the diagrams become so complicated when further divisions are made. Very soon the rule is introduced but probably not in the context of problems and so the child does not recognise the need for its use. Further there are several meanings attached to $\frac{1}{3} \times \frac{5}{9}$ – that of the area of a rectangle which has dimensions of less than one unit or one third part of a section of a pie or a series of multiplications and divisions within the set of whole numbers. We tend to assume that the child will acquire all the meanings although we may concentrate on only one in an introduction.

Fractions are not just an easy step from whole numbers. Their use introduces considerable problems for the child – new rules and new possibilities. Fractions are not used much in everyday life and a case must be made for their inclusion in the school mathematics curriculum. If the case is that the whole numbers have limitations (e.g. you cannot divide a smaller number by a larger) then the child must be at a stage where he sees the limitations.

The topic is taught in the primary school and again at different times in the secondary school but there may be a case to be made for a postponement of all work involving fractions until the secondary school stage. Partly re-teaching the topic throughout the secondary school without starting from the beginning (e.g. this is a half) causes frustration and boredom on the part of the learner.

6 Positive and negative numbers

The Positive and Negative Numbers test was devised by a student and even though the test had not been fully developed, it was included in the project's final round of large-scale testing. Although it has been decided not to publish the test, the results on some of the items still give useful insights into children's understanding of positive and negative numbers and these will be discussed in the present chapter.

These items can usefully be classified into a number of levels but first children's understanding will be described in relation to the separate operations of addition, subtraction and multiplication (there was only one division item). For convenience, the facilities quoted will be for the 14-year-old sample which consisted of 302 children. In addition 334 13-year-old and 182 15-year-old children took the test and a comparison of their performance is given at the end of the chapter.

Addition, subtraction and multiplication

Addition

As can be seen from Table 6.1 the addition items were generally answered correctly by well over 80 per cent of the 14-year-old pupils, even when one of the integers was negative (as in A4 onwards). The results suggest that most pupils could

Table 6.1   Addition items

Level	Percentage correct		Item	Correct answer	Dominant wrong answer (percentage)
1	97	A1.	$^+2 + {^+6} =$	$^+8$	
	94	A2.	$^+3 + \quad = {^+8}$		
	93	A3.	$^+5 + \quad = {^+7}$		
2	91	A4.	$^+5 + {^-5} =$	0	$^\pm10$ (5)
	88	A5.	$^+5$ then $^-4 =$	$^+1$	$^-1$ (5)
	88	A6.	$^+6$ then $^-3 =$	$^+3$	$^-3$ (5)
	87	A7.	$^+8 + {^-4} =$	$^+4$	$^-4$ (6)
	87	A8.	2 steps forwards and then 3 steps backwards is the same as....		
	85	A9.	$^-3$ then $^+1 =$	$^-2$	$^\pm4$ (4)
	85	A10.	$^-8 + {^+3} =$	$^-5$	$^\pm11$ (6)
	82	A11.	$^+2$ then $^-3 =$	$^-1$	

successfully model the items in terms of 'moves' or 'shifts' along the number line. Thus there was very little difference in facility between an item like A8 that involved shifts and required no knowledge of integers ('2 steps forwards and then 3 steps backwards is the same as . . . .') and a purely numerical item like A10 ($^-8 + {}^+3 =$).

The pattern of wrong answers shown in the Table suggests that some pupils, instead of using shifts, tried to solve the items by means of a rule which seemed to consist of ignoring the sign of the first integer and then adding the numerals if the second integer was positive ($+^+ \rightarrow +$) and subtracting if it was negative ($+^- \rightarrow -$).

Items with negative answers (A8, A9, A10, A11) tended to be more difficult than those with positive answers although the difference was not very great except perhaps in the case of item A11. Here the shifts cross the origin, which may have been ignored.

Subtraction

Some of the subtraction items are shown in Table 6.2. These items are more difficult to model, and though various methods of solution were suggested in the test itself, there is very little evidence to indicate that the children took notice of them. Instead, it appeared that most children tried to solve the items by making use of, or inventing, *rules*. However, these were usually inadequate and frequently changed from one item-type to another.

Table 6.2   Subtraction items

Level	Percentage correct	Item		Correct answer	Dominant wrong answer (percentage)	
2	84	S1.	$^+6 - {}^+6 =$	$\dot{0}$	$^\pm12$	(5)
3	77	S2.	$^+8 - {}^-6 =$	$+14$	$^\pm2$	(13)
	70	S3.	$^+6 - {}^+8 =$	-2	$^+2$	(14)
4	44	S4.	$^-2 - {}^-5 =$	$+3$	$^\pm7$	(37)
	36	S5.	$^-6 - {}^+3 =$	-9	$^\pm3$	(47)

The pattern of wrong answers is illuminating. For items that involved subtracting a positive integer (S1, S3, S5) the dominant strategy seemed to have been simply to subtract the numerals, followed by an attempt to determine the sign of the answer. This strategy is adequate for S1 (84 per cent), $6 - 6$ is 0 and the sign of 0 does not matter, but not quite sufficient for S3 (here 84 per cent again arrived at the correct numeral, 6, $8 \rightarrow 2$, but 14 per cent subsequently chose $^+2$ instead of $^-2$). For S5 the strategy produces the wrong numeral, 3 (47 per cent), and here less than half the pupils were successful.

For items that involved subtracting a negative integer (S2, S4) most pupils seem to have applied (or mis-applied) some version of the rule 'two minuses make a plus'. This strategy works for S2 ($^+8 - {}^-6 \rightarrow 8 + 6 \rightarrow 14$) but not for S4 where the addition of 2 and 5 produces the wrong numeral 7 (37 per cent). Here, as with S5, less than half the pupils succeeded with the item and it would seem that of the 5 items shown only these two test a clear understanding of subtraction.

Multiplication

It is probably even more difficult to model multiplication than it is subtraction of integers. However, the rule 'two minuses make a plus' can now be applied quite unambiguously and it seems likely that this is how the items in Table 6.3 were solved by most pupils. This is supported by the fact that these items proved to be surprisingly easy, with facilities similar to the subtraction item S2 (77 per cent) ($^+8 - ^-6 =$) which can be answered correctly using the same rule.

Table 6.3   Multiplication items

Level	Percentage correct	Item		Correct answer	Dominant wrong answer (percentage)	
3	80	M1.	$^-3 \times ^+4 =$	-12	$+12$	(10)
	79	M2.	$^+4 \times ^-4 =$	-16	$+16$	(7)
	78	M3.	$^+5 \times ^-2 =$	-10	$+10$	(11)
	76	M4.	$^-4 \times ^-2 =$	$+8$	-8	(12)

Levels of understanding

The items discussed in the previous section fell into four levels which are described below.

Level 1

The items at this level were the three addition items that involved only positive numbers (A1, A2, A3) and could therefore be solved by simply treating the given numbers as natural numbers. However, children at this level were able to cope with integers to the extent of regarding them as points on a number line, as in the items shown below which were answered correctly by 95 per cent and 94 per cent of the 14-year-old children respectively.

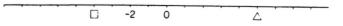

What number would go in △.
What number would go in □.

Level 2

The items classified at Level 2 were the remaining addition items (A4 to A11) and the subtraction item S1. These addition items involved negative as well as positive numbers but could be solved by regarding the numbers as a succession of shifts along the number line, which seemed to provide a highly effective concrete model.

In S1 ($^+6 --^+6 =$) children had only to recognise that the difference between two identical elements is zero, irrespective of the model used for the elements and the operation.

Level 3

The Level 3 items consisted of all the multiplication items and two subtraction items S2 and S3. With the exception of S3 it was possible to avoid constructing any meaning for the negative numbers involved by simply applying the rule 'two minuses make a plus'. In S3 ($^+6 - ^+8 =$), though the given numbers are straightforward (both positive), children had to be willing to go into the domain of negative numbers rather than simply take the smaller number from the larger, which is the strategy commonly used by younger children (see Chapter 3).

Level 4

The difference in facility between the items at Levels 3 and 4 was very much larger than the differences between the previous levels. The items classified at Level 4 were the two subtraction items S4 (44 per cent), $^-2 - ^-5 =$, and S5 (36 per cent), $^-6 - ^+3 =$, neither of which can be solved by a single-step rule as in Level 3, nor by means of a simple model like that applied in the addition items at Level 2.

Changes in performance by year

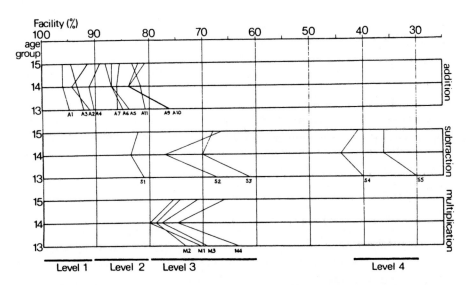

The diagram above gives the range of the four levels and shows the facilities of the items for the three age groups (13, 14, 15). Perhaps the most obvious feature of the diagram is that the facilities for the oldest children are in most cases lower than those for the 14 year olds. The reasons for this fall-off in performance are far from clear, though it may well in part be due to the older children having forgotten the meanings that were given to the integers when they were first introduced — which was probably several years earlier. As a result such children are likely to operate on the integers solely by means of rules which will be difficult to check for consistency if the rules lack meaningful support.

Implications for teaching

The various approaches to teaching integers can be classified into two main types, one of which is essentially abstract while the other relies on the use of concrete models or embodiments to give meaning to the integers and the operations to be performed upon them.

The advocates of the first approach argue that integers are abstract mathematical entities which form a self-contained system defined by certain rules, and which therefore should be taught as such. In particular, since the integers are not simply an extension of the natural numbers, they should not be presented as if they were. This approach, though uncompromising, has one clear advantage from a teaching point of view in that it avoids the difficult issue of trying to construct a unified model for the different operations on the integers. However, despite this advantage the results of the present research make it extremely doubtful whether the approach would be accessible to all but a minority of secondary school children (effectively, those who can cope with the Level 4 items). This conclusion is also supported by the work of Collis (1975, 1978) who has shown that the ability to work consistently within an abstract mathematical system requires formal operational thought. This is not to argue that most secondary school children are unwilling to work with rules as such, however arbitrary. Indeed it is clear from the experience of teaching that many children are only too happy to use or invent any number of rules for operating on the integers. The difficulty stems from the need to work consistently with such rules without recourse to an external, concrete referent, and it is this that most secondary school children seem unable to do.

As far as the second approach is concerned the results reported in this chapter are far more encouraging, though perhaps only within certain limits. Thus virtually all the children tested could use integers as points on a number line and the majority could cope with the additon of integers. However, even though the majority were also able to work within the domain of integers in the case of a fairly simple subtraction item like S3 ($^+6 - {}^+8 =$), the results also indicate that most secondary school children have a very limited understanding of subtraction (and of multiplication, though here children often succeeded by using a simple rule).

The contrast between children's understanding of addition and subtraction may stem in part from the models based on the number line that are commonly used to embody both operations.

$$a + b = c \qquad\qquad c - b = a$$

For addition the model is extremely straightforward and effective in that the result can be seen as a shift equivalent to a sequence of shifts. However, for subtraction, although the result is again a shift, the given integers are more likely to be seen as points. This makes the model far more difficult to use, not only because the operation is not seen as a simple sequence but also because the meanings given to the integers differ and are not consistent with the single meaning used for addition.

This change in meaning suggests that the number line should be abandoned, despite its proven effectiveness for addition, in favour of a more consistent approach, for example one in which the integers are regarded as discrete entities or objects, constructed in such a way that the positive integers cancel out the negative integers. The clear advantage of such a model is that the same meaning can be used for the integers both within and across the operations of additon and subtraction, and it seems likely that this would enhance children's understanding of subtraction in particular. However, the hopes placed in such a model should not be exaggerated. It remains to be seen whether the likely improvement in children's understanding of subtraction would be substantial since the coordinations involved would still be more complex than those required for addition. Also the model is as inappropriate for multiplication as the number line, but it is difficult to see a genuine way round this limitation without using an entirely abstract approach which, it has already been argued, is likely to leave most children with no understanding of integers at all.

7 Ratio and proportion

The CSMS ratio test was written to provide items involving both ratio and proportion. The items sometimes required the child to draw, sometimes to use numbers in relation to diagrams and at other times to use numbers presented in problems. Various ratios were used in different items. The examples were worded in such a way that technical terms like 'proportion', 'ratio' and 'similar' were avoided. In their final form the items mirrored aspects of ratio and proportion commonly taught in the secondary school with the exclusion of fractions, similar triangles and trigonometry. Fractions formed rather too large a segment to be part of the ratio paper and so merited a separate investigation; knowledge of similar triangles (although important) seemed to be very dependent on whether the topic had just been specifically taught and trigonometry was deemed to be too difficult for the younger children being tested.

It was hoped that a purely mechanical application of a technique would not be encouraged; thus, for example, it was difficult to enlarge the figure on the test paper using a centre of enlargement. However, children who were sufficiently comfortable with a 'rule' or 'technique' to see its applicability within a problem could of course have applied it; those who could only use a rule when the question presented was in a particular form were at a disadvantage and normally resorted to other methods.

Topics covered

An investigation of British Secondary School Mathematics textbooks resulted in the following list of aspects of ratio which appear germane to the subject:

(1) Doubling or halving
(2) Multiplication by an integer
(3) Given a rate per unit, apply this rate
(4) Find a rate per unit and then apply it
(5) Enlargement of a drawing in ratio $2:1$
(6) Enlargement of a drawing in ratio $3:2$
(7) Enlargement of a drawing in ratio $5:3$ etc.
(8) Finding a ratio $a:b$ using an intermediate quantity, c, i.e. given relationships a to c, b to c
(9) Using a fractional multiplier
(10) Simple percentages.

Methods used by the children to solve ratio and proportion questions

The methods the children used are described with illustrations from the items and the interviews. An immediate feature apparent in the interviews was that only *one* child quoted the rule or formula $\frac{a}{b} = \frac{c}{d}$ (three values known, one must find the fourth). No child actually quoted the unitary method, e.g. given that 14 metres of calico cost 63p find the price of 24 metres.

$$14\,\text{m calico cost} \quad 63\text{p}$$

$$1\,\text{m calico cost} \quad \frac{63}{14}\text{p}$$

$$24\,\text{m calico cost} \quad \frac{63 \times 24}{14}\text{p}$$

Some children did essentially seek a relationship which involved $a:1$ but did not provide explanations in the above form. Of the 2257 children tested in 1976 only 20 wrote down an equation of the form $\frac{a}{b} = \frac{c}{d}$ and used it consistently and correctly. Fifteen of this 20 were from the same school but they were the high attainers out of the 100 from the school who had been tested. There was little evidence that the taught rule $\frac{a}{b} = \frac{c}{d}$ was remembered and used by the children. In fact most children on interview changed the method they used continuously, adapting to what they saw as the demand of the question. Generally they avoided multiplying by a fraction and tended to build up an answer in small segments, adding them together at the end. Sometimes this meant they were carrying a considerable amount of information in their heads and were unable to recall all the various segments and thus gave an incorrect answer. Often the children used a method of 'building up' that was specific to the problem posed; in, for example, the eel question that follows later (eels are fed in proportion to their length) they reasoned that if the food for a 10 cm eel is added to the food for a 15 cm eel then you have the food for a 25 cm eel.

Doubling and halving

Doubling and halving are the easiest aspects of ratio, when presented in either problem form or drawing. When given a line segment and its enlargement and then asked to complete an open rectilinear figure like this:

72 per cent of the children could recognise the ratio 2:1 and successfully double the upright. Many forgot to enlarge the gaps so that the resulting enlargement

did not look like the original. In a problem involving a recipe for eight people which asked for the ingredients for four, most children halved; very few found how much for one person and then multiplied. The success of a doubling approach is no indication of what happens when the ratio is not $2:1$. Some children in fact view all enlargements as a requirement to double and all reduction as a requirement to halve.

Building up using halving

The recipe question, besides asking for amounts of ingredients for four people (given the amounts for eight), then went on to ask for the correct amounts for six people. Most children used the information they had obtained for four and then said 'another two people is half of four people' and thus obtained two numbers and added them. This method was successful as long as the resultants were whole numbers or a half but became very cumbersome when a fraction was involved from the outset. One of the ingredients was $\frac{1}{2}$ pint of cream for eight people. Thus for six people (using this building up method) the children had to compute $(\frac{1}{2}$ of $\frac{1}{2})$ $+ \frac{1}{2}(\frac{1}{2}$ of $\frac{1}{2})$. This proved to be too difficult and about 20 per cent in each year group of the large sample gave the answer $\frac{1}{3}$. They were seeking a fraction between $\frac{1}{2}$ and $\frac{1}{4}$ but chose $\frac{1}{3}$. The difference in facility between the same question asked of whole number amounts and the fractional amount was 60 per cent (80 + per cent for whole numbers and 20 + per cent for the fraction). It was noticeable that the children when interviewed did not multiply by a fraction.

Other methods used to build up to an answer

To illustrate the diversity of 'building up' methods we will look at two examples taken and adapted from Piaget's *Epistemologie et Psychologie de la Fonction* (1968). Piaget describes the true understanding of proportion as a formal operational task. Piaget's eel question is concerned with the amount of food given to eels of different lengths, the amount being proportionate to the length of the eel. The question is in two forms; one where the food is made up of discrete objects and the other where it is formed from a continuous quantity (e.g. length of biscuit). In the original experiment the eel lengths are in the ratio $1:2:3$. The CSMS test version had two forms:

(a)   discrete objects and eels in ratio $1:2:3$ and
(b)   continuous quantity and eels in ratio $2:3:5$.

2a.   There are 3 eels A, B and C in the tank at the Zoo.

The eels are fed sprats, the number depending on their length. If C is fed two sprats, how many sprats should (i) B and (ii) A be fed to match?

(iii)   If B eats 12 sprats, how many sprats should A be fed to match?
(iv)   If A gets 9 sprats, how many sprats should B get to match?

2b.   Three other eels, X, Y and Z are fed with fishfingers, the length of the fishfinger depending on the length of the eel.

25 cm long	Z

15 cm long	Y

10 cm long	X

(i)   If X has a fishfinger 2 cm long, how long should the fishfinger given to Z be?
(ii)   If Y has a fishfinger 9 cm long, how long should the fishfinger given to Z be?
If Z has a fishfinger 10 cm long, how long should the fishfingers given to (iii) X and (iv) Y be?

The Piagetian question had been made more difficult by the substitution in the second part of eel lengths which were not in the ratio $1:2:3$. Only two children out of 29 interviewed used the same method throughout all parts of the eel question. 2a (i) and (ii), where the amount of food given to the smallest eel was stated, were very often solved by simply doubling and trebling. 'C is two, times it by 2, and then times it by three'. Other children however did not multiply but built up to the answer 'Its 2, 4, 6 . . . Five, ten, fifteen goes up in fives. It's five up to there, add another five which is two, add another five – its another two, so 2, 4, 6'. These two methods might be regarded as the finding of a rate per 5 cm of eel. Sometimes the method degenerated into a number pattern answer, e.g. 'It just goes down in steps of two'.

When the question required operations more complicated than doubling or trebling, the number of methods used by the children increased. In Question 2b multiplication by a fraction was infrequent, two children multiplied by $2\frac{1}{2}$ when given the amount of food for the 10 cm eel and asked for the amount eaten by the 25 cm eel. Faced with 'The 15 cm eel has 9 cm food, how much food should be given to the 25 cm eel?', no child multiplied by $\frac{5}{3}$. Building up to an answer was applied in a number of ways, each of which entailed finding parts of the answer and finally adding them together. For example, instead of multiplying by $\frac{5}{3}$ two children used this argument:

10 is $\frac{2}{3}$ of 15. Two thirds of 9 is 6. That is $15 + 10$ so add those two amounts.

Finding a rate and then building up can be illustrated from 2b (i) (2 cm food for 10 cm eel, how much for the 25 cm eel?).

    (i)   10 gets 2 cm. 15 is half again. The amount 15 gets plus the amount that 10 gets is the amount 25 gets.

   (ii)   The answer is five, two of those (X) would make twenty and then another 5 cm which is 1 cm fishfinger.

In question 2b, finding the rate per 5 cm eel and multiplying by an integer was an infrequently used strategy, unlike 2a where the rate per 5 cm was given and only doubling and trebling was needed. Finding the rate may have been the difficulty, e.g. Mary faced with 9 cm food for 15 cm eel became thoroughly confused by the computation.

    Y has 9 cm. Divide 9 into 15, goes once with 4 left over. 1.4 into 10 goes 7, it's 9.8. So it's 7.2 cm long.

Two levels of answer described by Piaget (1968) were level II, adding on one for a larger eel and level III, adding on a fixed amount greater than one (we looked at the addition of two). Thus in Question 2a(iii) the child at level II would reason that A is longer than B, B has 12 sprats so A has 13 sprats. By adding two the child would give 14 sprats to A. Piaget's level II replies occurred very infrequently in the large sample (n = 2257). The highest incidence in the eel question was three per cent on Question 2b(i). 'Adding on 2' replies were given by seven per cent on 2a(iv) and by between nine per cent and 13 per cent on Question 2b(i). Some children doubled or halved when the ratio was not 2:1. For example on Question 2b:

Child	It's four and eight. That is double the two makes 15.
Interviewer	Why did you double. Y is not double X's length is it?
C	No.
I	Would doubling make it fair?
C	Yes.

Doubling and halving when inappropriate occurred at the following levels on the eel questions 2b(iii) and 2b(iv):

Table 7.1   Percentage of Children Doubling on Eel questions
(n = 2257, age 13–15)

	13 yr	14 yr	15 yr
2b(iii)	11.7	11.6	8.8
2b(iv)	20.0	16.6	12.9

The complexity of the task in Question 2a is much increased when the amount given is for the eel of middle length and not the smallest. The questions in item 2a are however much easier than those in item 2b where the ratios are not x to one, where x is an integer.

    Incorrect methods sometimes, of course, give the correct answer. A child who views all enlargement as doubling will of course be correct when the ratio is 2:1.

Similarly, a number pattern used in 2a(i) and 2a(ii) may lead one to suppose that the child is applying a ratio. Simple addition of two to each of the lengths would give the correct answer in 2a(i) and 2a(ii) so that a child who finds it works once might be tempted to continue using it.

Naive methods

The most unsophisticated and naive answers met on interview were those where the child knew that a figure was to be enlarged but did not use the data or even measure. Thus asked to enlarge a rectangle the child would draw any rectangle or enlarge the length but not the height. The idea that the new figure should be *similar* to the original was missing. Indeed the word 'similar' was not used in the

interviews or test paper since to most children it meant 'vaguely the same shape'. The 'same shape' is also a difficult notion when dealing with rectilinear figures since all triangles are the 'same shape' in that they are all triangles. With non-rectilinear figures 'the same shape but bigger' seemed to convey what was needed.

An example of a primitive method on the eel question, 'If X has a fishfinger 2 cm long how long should the fishfingers given to Y and Z be?' is:

James   Five and eight. It just goes up more and more.
Interviewer   Why did you decide two, five, eight?
J   Well, they're ten, 15, 25 − it doesn't go to 20. So guess it goes up more.
I   Why five and not six for Y.
J   Suppose it could do but five goes into that and that. You could have had ten.
I   You could have had ten but you chose eight.
J   They're more equal but either would be alright.

Addition strategy

The addition of a small amount to make an enlargement has already been mentioned in the discussion of Piaget's eels. A much more sophisticated form of addition is that where the child uses the data in the question but concentrates on $a - b$ not $\frac{a}{b}$. Piaget describes this as a typical answer from a child at the late concrete level. Karplus sees it as not part of a sequence of understanding but a method which should be corrected. The question adapted from Karplus (1975) and used on the CSMS test was:

5.

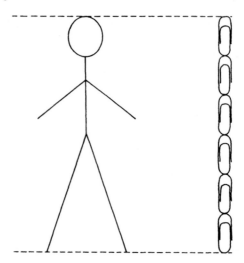

You can see the height of Mr. Short measured with paper clips. Mr. Short has a friend Mr. Tall. When we measure their heights with matchsticks:

> Mr. Short's height is four matchsticks
> Mr. Tall's height is six matchsticks

How many paperclips are needed for Mr. Tall's height?

The incorrect strategy in which the child concentrated on the difference $a - b$ rather than $\frac{a}{b}$ would result in:

> Mr Short needs two more paperclips than matchsticks so Mr. Tall needs two more paperclips than matchsticks, so the answer is eight.

The use of this 'addition' strategy occurred at the 25–50 per cent level on four of the most difficult items on the ratio test.

The Karplus question was one of the four. The other three involved similar figures in which the child had to recognise that the question involved ratio and then find the enlargement factor. The need for an enlargement was recognised but that a *multiplicative* strategy was needed did not seem obvious.

All four were given in diagrammatic form. Two of the questions needed a ratio of 3:2, the other (the most difficult) required a ratio of 5:3. From the total sample of 2257, 30 per cent of the children consistently used the addition strategy on at least three of the four questions mentioned. Defining these children as 'adders' we looked at their overall performance. These 'adders' could cope with the questions which involved halving or doubling (including the eel questions, 2a(i) and (ii)). About half of them could successfully solve items 2a(iii), 2a(iv) and 2b(i) but this dropped to 20 per cent on items 2b(ii), (iii) and (iv), that is where the building up method using wholes and halves could not be so easily applied. On items 2a(ii), (iii) and 2b(i) some 12–16 per cent of the adders were 'adding on two' as described above. On items 2b(ii), (iii) and (iv) only 6–8 per cent of them were using this strategy but the success rate had fallen to 20 per cent. The addition strategy was not confined to one age group and although more 13 year olds than 15 year olds used it, the frequency of use never differed by more than seven per cent.

The method is obviously not used by the least able; it has its attractions since the child feels that he has 'done something'. Since his methods on the easier questions usually involved an addition of segments of an answer, addition on the harder items would seem logical. Indeed when asked to enlarge a rectangle

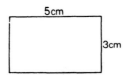

so that the new base was 12, the height using the addition strategy would be 10 cm. This amount makes the Figure look like a square, and so produced an obvious distortion. Some children aware that it was too large fell back on the method, 'take two 3 cm pieces and the extra two, answer 8 cm'.

Methods when parts of the data were omitted

An example when $a:d$ had to be found and the question provided an intermediate quantity c was:

6.   In a particular metal alloy there are
       1 part mercury to 5 parts copper
       3 parts tin to 10 parts copper
       8 parts zinc to 15 parts copper
You would need how many parts mercury to how many parts tin?
You would need how many parts zinc to how many parts tin?

The first part was easier than the second since if one noticed that the copper had to be used the ratio was then 2:1. The most common error was that of ignoring the copper altogether, e.g.

Child   One part mercury to three parts tin
Interviewer   Why did you decide on those
C   It's there, one part mercury and three parts tin
I   Doesn't it make any difference that I have five parts copper and ten parts copper?
C   Yes, it does I suppose
I   What are you going to do about it?
C   Well you can't work it out like that
Part B
C   Eight parts zinc to three parts tin
I   Does it matter that you have ten parts copper and 15 parts copper?
C   No.

In the large sample 20 per cent of each year group gave the answers 1:3 or 8:3. A lesser number simply doubled both mercury and tin amounts and felt they had 'done something' by producing 2:6. Only one child on interview tried to obtain a uniform amount of copper (30 parts). Those who were successful tended to use a 'building up' method, e.g. 'Five parts more copper would give me one and a half parts more tin so $(3 + 1\frac{1}{2})$ tin to eight zinc'.

Percentage questions

The questions on percentage used on the CSMS test explained the meaning of per cent in terms of 'per 100'. Each item involved the use of a rate per 100, e.g. '24 out of 800 cars are faulty, what percentage is this?' Most children found these items very difficult and tended to use a misremembered rule or concentrated on just one word in the problem, e.g. 'reduced'. Three of the percentage questions and the overall results are shown below.

8.   % means *per cent* or *per 100* so 3 % is 3 out of every 100.
b.   6% of children in a school have free dinners. There are 250 children in the school. How many children have free dinners?

Age	13	14	15
Success rate	36.6	45.5	57.7 per cent

c.   The newspaper says that 24 out of 800 Avenger cars have a faulty engine. What percentage is this?

Age	13	14	15
Success rate	32.0	39.6	47.6 per cent

d.   The price of a coat is £20. In the sale it is reduced by 5 %. How much does it now cost?

Age	13	14	15
Success rate	20.0	26.5	35.4 per cent

The crucial word in Question 8d seems to have been 'reduced'. Two different methods of achieving the reduction gave rise to either of the incorrect answers 15 or 16. Fifteen was presumably obtained by working $20 - 5$ ignoring the symbols £ and %. Sixteen can be obtained from $20 - \frac{20}{5}$ again ignoring £ and %.

Taking the answers of 15 and 16 together, their incidence was high as indicated by the following figures: 13 yr $- 40.9$ per cent; 14 yr $- 35.9$ per cent; 15 yr $- 27.4$ per cent.

The belief by some children that when faced with a percentage question one always divides by 100 was reflected by the answers to question 8c. Here a popular answer was 192, presumably obtained by $\frac{800}{100} \times 24 = 192$. This occurred as follows: 13 yr $- 7.4$ per cent; 14 yr $- 10.4$ per cent; 15 yr $- 8.8$ per cent.

Summary

The methods the children use for solving ratio and proportion problems do not involve the rule $\frac{a}{b} = \frac{c}{d}$, nor do they involve a single multiplication by a fraction. Instead there is a constant effort to obtain an answer by using addition in some form or other. This leads to considerable difficulty when the problem is to 'make smaller', e.g. the eel question, 2b(iv), since the process has to be reversed and some form of subtraction used. The method also leads to difficulties when the segments in the building up process are not easily obtained by taking the amount once or twice and adding, e.g. using 5:3 in the following example:

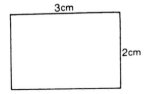

Complete this diagram so it is the same shape but larger than the one above

5 cm

Here the building up method requires one to say 'take two and a bit of two for the new upright'. Since the fraction of two to be added is not half of it the child has difficulty and often opts for either three or two and a half for the upright.

The common incorrect 'addition strategy' shows a continued desire to see enlargement as a process of adding not multiplying. When enlargement involves the drawing of figures children are very often unable to compare the shape of the new figure with that of the original and they produce figures which show considerable dissimilarity.

Results of the survey

The facilities of the items

There were 27 items on the ratio and proportion test and the children took approximately 30 minutes to complete the test. Obviously not all aspects of the topic could be covered in a test of this length so there was an absence of items in certain facility ranges when the results were analysed. The facilities ranged from 95 per cent to 12.5 per cent (total for all three years). In Table 7.2 the facilities of items already mentioned are shown and are for the three years together (2257 children).

Groups and levels of understanding

The method of analysis used on the data in order to obtain groups of items has been explained in Chapter one. Using the data from testing with the ratio paper four groups of items were formed, each containing four of five items. The facility range for each is shown in Table 7.3.

The very easiest items dealing with ratio and proportion were not associated highly and tended to form two pairs (two recipe questions and two eel questions). However, in order to form a starting level for the topic they have been grouped together and are described as Level 1.

Table 7.2   Facilities of items on ratio paper 1976 (items already discussed in text)

% success		
100		
	95 %	Recipe question given 8, find 4 people's amounts (whole numbers)
90		
	85 %	Recipe question, how much for 6, answer $1\frac{1}{2}$
		Doubling on eel question 2a(i)
80		
	78 %	Recipe question, how much for 6 (whole numbers only)
	72 %	Trebling on eel question 2a(ii)
		Doubling length of line segment
70		
60		
50	50 %	Eel questions 2a(iii), 2a(iv), 2b(i)
		Another question involving sharing in ratio $1:2:3$
	45 %	Percentage question 8b
40		
	39 %	Percentage question 8c
	32 %	Mr. Short and Mr. Tall; Chemical compounds (question 6, pt. 1)
30		
	30−27 %	Eel questions 2b(ii), (iii), (iv); Percentage question 8d
	26 %	Cream question in recipe
20		
	20 %	Enlargement non rectilinear figures, ratio $3:2$
	13−12 %	Chemical compounds pt. 2 and enlarging line segment in ratio $5:3$
10		

Table 7.3   Levels of understanding in ratio and proportion

	Facility range	Pass mark
Level 1	79−95 per cent	3/5
Level 2	39−51 per cent	3/5
Level 3	27−33 per cent	4/6
Level 4	12.5−21 per cent	3/4

Table 7.4   Description of levels

Level	Description	Type of item (from those discussed in text)
0	Unable to make a coherent attempt at any of the questions	
1	No rate needed or rate given. Multiplication by 2, 3 or taking half	Recipe question — given ingredients for 8 find the amounts for 4 Eel question 2a(i), (ii)
2	Rate easy to find or answer can be obtained by taking an amount then half as much again	Eel question 2a(iii), (iv) Percentage question 8c
3	Rate must be found and is harder to find than above. Fraction operation also in this group.	Mr. Short and Mr. Tall Cream question in recipe
4	Must recognise that ratio is needed, the questions are complex in either numbers needed or setting	Enlargement in ratio 5 : 3

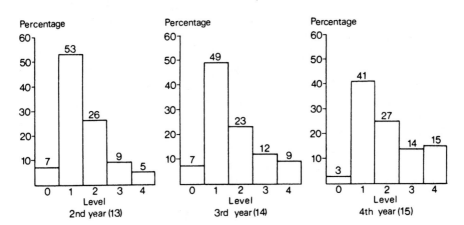

Fig. 7.1   Percentage of children who achieved each level for the three year groups in 1976

Comparison of performance by year

The fourth year had more children achieving Level 4 and less at Level 0 than the two younger age groups. On all items the fourth year scored best but sometimes only just. The very easiest items were equally well done by all the age groups. The hardest items tended to have little difference in facility for the second and third years but sometimes as much as a 12 per cent increase in facility for the fourth years. The percentage questions showed an eight per cent difference between years, the same pattern being shown on the harder eel questions. On no question was the fourth year's score more than 14 per cent higher than that of a younger age group. There is a very slow progression of attainment from year to year. This is clearly shown in the results of a longitudinal survey involving about 100 children which is reported in Chapter 12.

Implications for teaching

Methods used by children

The methods used by children to solve problems in ratio and proportion vary with the problem presented. One of the most popular methods is that of *building up* to an answer by dealing with small segments of the problem and then *adding* together the answers. This is adequate for items involving $2:1$, $3:2$ or even $5:2$ but when one can no longer say 'take the amount, take it again and take half of it' the ensuing computation defeats the child and he resorts very often to an incorrect method. Level 2 questions can be solved by this building-up method using addition. Level 3 questions are more complex and the use of this strategy is cumbersome, e.g. in the question '$\frac{1}{2}$ pint of cream for 8 people, how much for 6 people?' the method results in $(\frac{1}{2}$ of $\frac{1}{2}) + \frac{1}{2}(\frac{1}{2}$ of $\frac{1}{2})$. The fact that most children do not progress beyond Level 2 items shows that they probably see ratio as an *additive* operation and essentially replace multiplication by repeated addition. When this is no longer possible one popular incorrect method is that of considering the difference $a - b$ rather than the ratio $a:b$, the strategy is additive but not in the same sense as the 'building up' described above. The use of this method is widespread and is used by 30 per cent of the population on the hardest items. Since it is so prevalent teachers should be aware of its existence and try to eradicate it. There is evidence (Kuntz and Karplus 1977, Renner and Paske 1977) to show that children who are presented with practical problems (e.g. gears) which need a ratio for solution, do improve and abandon the 'addition strategy'. Another possible eradication strategy is that of showing gross distortions that occur when the difference $a - b$ is used. In problems where the $a - b$ method gives very nearly the same answer as the correct method, the addition strategy is only too plausible.

Similar figures

Similar triangles appear early in the introduction to ratio in most secondary textbooks and children are expected to recognise when triangles are similar to each other and the properties they possess. On interview it was found that the word 'similar' had little meaning for many children. In everyday language the word is used in a non-technical sense to mean 'approximately the same'. There was particular difficulty with the word when similar triangles or rectangles were under discussion. The CSMS test items dealing with similar figures (not triangles or rectangles) were some of the hardest on the test paper. Using ratio to share amounts between people 'so that it is fair' seemed to be much easier than dealing with a comparison of two figures. In enlarging figures there is the danger of being so engrossed in the method to be used that the child ignores the fact that the resulting enlargement should be the same shape as the original.

Progression year to year

The majority of children do not progress beyond doubling, halving and using doubles and halves to 'build up' to an answer. This ability is no guide to how the child would tackle a ratio of say $5:3$. Although progress is made from year to year it is generally only a progression of one level. There is no rapid progress over three years to the ability to handle all ratios. The ability to handle all ratios necessitates the use of *multiplication* by a fraction. Teachers should be aware that children avoid multiplying by a fraction and use repeated addition. The gradual introduction of examples which need multiplication should be based firmly on the knowledge of what the children do most naturally. We cannot expect them to abandon a method which gives success because we state a new method is available. The knowledge of the methods used by children is particularly relevant when the teacher is correcting errors since if the correction takes the form of a repetition of the 'teacher method', the child's error made in a 'child method' is not touched.

Examples given to children should be graded very carefully — it is of little value to have doubling and $5:3$ in the same introductory exercise. The introduction of non-whole numbers into a problem does not make the question a little harder but a lot harder.

Algorithms

Finally, teaching an algorithm such as $\frac{a}{b} = \frac{c}{d}$ is of little value unless the child understands the need for it and is capable of using it. Children who are not at a level suitable to the understanding of $\frac{a}{b} = \frac{c}{d}$ will just forget the formula.

8   Algebra

Introduction

In devising the algebra test, an attempt has been made to sample a wide range of typical secondary school algebra (or, more strictly, generalised arithmetic) activities, including procedures like substituting and simplifying, and tasks such as constructing, interpreting and solving equations. However, it soon became apparent during the development of the test that such activities, singly and in combination, were far too numerous to be investigated thoroughly by one test, and that other, more fundamental criteria had to be found for determining the test's specific content and for making sense of the results. Eventually just two criteria were chosen, which might briefly be called the structural complexity of the items, and the meaning that can be given to the letters. The nature of these criteria, and the way they emerged, are described in the following section, which shows how they apply to one of the questions from the test.

Criteria for constructing and analysing the test

The Table below shows the responses that were regarded as correct for Question 9 of the algebra test (in which children were asked to write down an expression for the perimeter of each shape). For brevity, the percentages here, and in much of

Table 8.1   Correct responses to Question 9 (14 year olds)

9(i)	9(ii)	9(iii)	9(iv)
			 Part of this Figure is not drawn. There are n sides altogether all of length 2
3e 94%	4h + t 68% 4h + 1t	2u + 16 64% u + u + 16 2u + 25 + 16	2n 38% n2

what follows, are just for the 14-year-old sample (which consisted of just under 1000 children).

The percentages (facilities) in the table show that the four items are progressively more difficult, and whilst nearly all the (14-year-old) children could cope with 9(i), less than half answered 9(iv) correctly. Collis (1975) and Halford (1978) have shown that an important determinant of facility is the 'structural complexity' of an item, which, for Question 9, might be characterised by the number of 'variables' that the items involve. This criterion accounts very effectively for the difference in facility of the first two items, but it is clearly not sufficient for some of the other item pairs. For example, it might have been expected that 9(iii) (one 'variable') would be easier than 9(ii) (two 'variables') and that 9(i) and 9(iv) would be much closer in facility, in as much as their answers are of the same form.

Fortunately, a way of overcoming this deficiency is also provided by Collis, who has shown that the *nature* of the elements in an item (as well as its structural complexity) can have a marked effect on facility. The main distinction that Collis makes is whether the elements are small numbers, large numbers or letters. But more important than these specific categories is Collis's argument that the difficulties stem from the extent to which the elements lack *meaning* for the child. For example, an eight-year-old child might be quite happy with an expression like 2 + 3, because he can relate the elements, and their combination, directly to his everyday reality (two marbles and three marbles put together make five marbles) but he may not yet accept that an expression involving numbers beyond his verifiable range, like 274 + 356, also gives a stable, unique result.

In generalised arithmetic this suggests that the extent to which the letters are meaningful to children will be of vital importance in determining item difficulty. Furthermore, it seems likely that children may give *different meanings* to the letters, which in turn would affect item difficulty in that some items might be solved in unexpected ways.

To see what some of these different meanings might be, and how this can help explain differences in facility, it is worth returning to Question 9, and examining, in particular, some of the inadequate answers that children gave, since these provide useful clues to how the items were tackled.

It was mentioned earlier than 9(i) and 9(iv) have answers of the same form (3e and 2n), yet 9(iv) is much more difficult. How, then, are the demands of the two items different? In 9(iv), n is fairly clearly defined as a number (we are told the figure has n sides) which has to be operated upon even though its value is not known. Collis has shown that many secondary school children find this extremely difficult. Faced with simultaneous equations, for example, he found that many children gave numerical values to the letters before manipulating them in any way. The same kind of thing seems to be happening here, in that a sizeable proportion of children (18 per cent) gave straight numerical answers like 32, 34, etc., which seem to have been arrived at by counting the sides actually drawn (of which there are 16 and a bit) or by literally closing the figure by adding a few more sides. Clearly, these children were not willing to use the letter as a numerical entity in its own right.

If this accounts for the difficulty of 9(iv) then at first sight it ought also to apply to 9(i) and the other two items as the letters here also represent unknown numbers. However, from the child's point of view, it can be argued that in these items the letters may stand for something much simpler, that is they can be regarded as just *names* or *labels* for the sides themselves (and *not* the unknown *lengths* of the sides) which then simply have to be collected together. This view is supported by the fact that many children (27 per cent) gave answers like 2u + 2.5 + 1.6 instead of 2u + 16 which is very much an act of collecting. Similarly, children wrote 4h + 1t rather than 4h + t (through a shortcoming in the marking-scheme the relative proportions here are not known) and some (20 per cent) just produced a list of elements giving answers like 4h, t or 4ht or even hhhht.

To sum up, the responses to Question 9 revealed three different ways in which the letters were interpreted or used. Less than half the children seemed able to use a letter as a numerical entity in its own right (this might be called a 'specific unknown'), instead the letter was 'evaluated' (16 or 17 sides rather than n sides), or regarded as an 'object' (the sides consisted of two u's, two 5's and a 6).

Children used the letters in these three ways in other questions on the algebra test, and altogether six different ways of interpreting and using the letters were identified. These are described briefly below, and a further discussion follows.

Children's interpretations of the letters

Letter evaluated

This category applies to responses where the letter is assigned a numerical value from the outset.

Letter not used

Here the children ignore the letter, or at best acknowledge its existence but without giving it a meaning.

Letter used as an object

The letter is regarded as a shorthand for an object or as an object in its own right.

Letter used as a specific unknown

Children regard a letter as a specific but unknown number, and can operate upon it directly.

Letter used as a generalised number

The letter is seen as representing, or at least as being able to take, several values rather than just one.

Letter used as a variable

The letter is seen as representing a range of unspecified values, and a systematic relationship is seen to exist between two such sets of values.

The interpretation that children chose to use was dependent in part on the nature of the question — not all categories were always relevant — and also on the question's complexity. Generally the first three categories indicate a low level of response, and it can be argued that for children to have any real understanding of even the beginnings of algebra they need to be able to cope with items that require the use of a letter as a specific unknown, at least when the structure of such items is simple. Most children could not do this consistently (13, 14 *or* 15 year olds) and used one of the first three interpretations instead. Very few children reached the high degree of understanding required to interpret a letter as a variable.

Apart from these six categories, children's responses and the items themselves, were classified into different 'levels of understanding' (in common with the other CSMS tests). Four levels were used. A full discussion of these levels, and their justification, is given on page 112, but some indication of how they relate to the six categories should become apparent from what follows, in which the meaning of each category is described in more detail.

Letter evaluated

This is one of the three interpretations by which children avoid having to operate on a specific unkown, in this case by simply giving the unknown a value. The category also refers to items where children are asked to *find* a specific value for an unknown, but again without first having to operate on the unknown. This applies to question 6(i), 11(i) and 11(ii) in Table 8.2 below but *not* to 14.

Table 8.2   Children's responses (14 year olds)

6(i) (Level 1)		11(i) (Level 2)		11(ii) (Level 2)		14 (Level 3)	
What can you say about a if $a + 5 = 8$		What can you say about u if $u = v + 3$ and $v = 1$		What can you say about m if $m = 3n + 1$ and $n = 4$		What can you say about r if $r = s + t$ and $r + s + t = 30$	
$a = 3$	92%	$u = 4$	61%	$m = 13$	62%	$r = 15$	35%
						$r = 30 - s - t$	6%
		$u = 2$	14%	other values	14%	$r = 10$	21%

As can be seen, 6(i) is answered correctly by nearly all the children (all they need to do is recall a familiar number-bond, or count on from five till they reach eight).

Both parts of question 11 are harder (Level 2) but are still answered correctly by a majority of the children. The increase in difficulty is probably mainly due to the fact that the items involve two unknowns rather than one which makes the first equation in each item 'ambiguous' (for example $u = v + 3$ is true for more than one pair of values). However, this ambiguity is resolved as soon as the second equation is reached ($v = 1$).

Question 14 is harder still (Level 3). It can be solved by replacing $s + t$ by r in the second equation, but this involves handling a letter which is still unknown (though r can now be evaluated from $r + r = 30$), which puts the item into the specific unknown category. The substantial number of children who wrote $r = 10$ seem to have avoided this higher category by evaluating r directly from the second equation $(10 + 10 + 10 = 30)$.

Letter not used

In Question 5 the first two parts, but not 5(iii), can both be solved by 'not using' the letters (see Table 8.3). 5(i) proved to be very easy (Level 1) even though it seems to involve two unknowns. However, nothing need to be done to these unknowns. They can essentially be eliminated by a matching procedure which focuses attention on $+ 2$, by which the left-hand side of the equations differ: $+ 2$ is then simply applied to 43.

Table 8.3   Children's responses (14 year olds)

5(i) (Level 1)		5(ii)		5(iii) (Level 3)	
If $a + b = 43$ $\quad a + b + 2 = \ldots$		If $n - 246 = 762$ $\quad n - 247 = \ldots$		If $e + f = 8$ $\quad e + f + g = \ldots$	
45	97%	761	74%	$8 + g$	41%
		763	13%	15	2%
		Other values	8%	12	26%
				8g	3%
				9	6%

The letter in 5(ii) can be avoided in the same kind of way by matching the two equations (although there was also evidence that some children first evaluated the letter, which tended to lead to arithmetical errors because the numbers were large). The item was more difficult than 5(i) because of the large numbers and because the operation (-1) was implicit and counter-intuitive. The fact that 247 is greater than 246 persuaded some children to add 1 to 762 instead of subtracting.

5(iii) can also be solved by matching, but though e and f can be avoided in this way, children still have to operate with g which means usii.g a specific unknown (this puts the item at Level 3). Many children tried to resolve this difficulty by *evaluating* g — often in a quite logical way — which led to answers like 12 $(4 + 4 + 4 = 12)$ or 15 (g is the seventh letter of the alphabet). Others just added 1, presumably because this was the simplest way of making the answer bigger.

Letter as object

This category has already been discussed in the context of the perimeters in Question 9, where, for the first three figures, the letters can be regarded as denoting the sides of the figures rather than their unknown lengths.

This interpretation of the letters can also be used successfully in some, but not all, of the parts of Question 13 illustrated below, in which children were asked to simplify various algebraic expressions.

Table 8.4   Correct responses (14 year olds)

13(i) (Level 1)		13(iv) (Level 2)		13(viii) (Level 3)		13(v) (Level 4)	
$2a + 5a =$		$2a + 5b + a =$		$3a - b + a =$		$(a - b) + b =$	
7a	86%	3a + 5b	60%	4a − b	47%	a	23%

13(i) and 13(iv) can be solved by thinking of the letter(s) as a shorthand for, say, apples (and bananas) so that $2a + 5b + a$ becomes 2 apples and 5 bananas and another apple, making 3 apples and 5 bananas in all.

Or the letters can be thought of as objects in their own right (2 a's, 5 b's and another a). This makes the items far easier than if the letters had to be treated as specific unknowns. However, both approaches begin to break down in 13(viii) (Level 3) — 3 apples take away one banana makes little immediate sense (unless there already are some bananas), nor does 3 a's take away one b unless b is thought of as a number. Exactly the same kind of difficulty occurs with $(a - b)$ in 13(v), and this difficulty is heightened by the brackets which deliberately focus attention on an expression which cannot itself be simplified. However it is not the brackets alone that are the problem — a similar item, but involving addition throughout, $(a + b) + a$, was far easier (53 per cent) and quite close in facility to 13(iv) which was without brackets.

Using a letter as an object, which amounts to reducing the letter's meaning from something quite abstract to something far more concrete and 'real', allowed many children to answer certain items successfully which they would not have coped with if they had had to use the intended meaning of the letter. However this reduction in meaning frequently occurred when it was *not* appropriate. This happened particularly with items that involved 'objects' (cabbages, pencils, wages, hours, etc.) but where it was essential to distinguish between the objects themselves and their *number*. This distinction can sometimes be very hard to grasp. A classic example is (or used to be) the translation of the relationship 'one shilling equals 12 pence' into 's = 12d', (letter as object) instead of 'd = 12s'.

This confusion arose in Question 22 (Table 8.5), even with children who did well on the test as a whole. To solve this item, the letters have to be regarded, at a

Table 8.5   Children's responses (14 year olds)

Question 22 (Level 4)

Blue pencils cost 5 pence each and red pencils cost 6 pence each. I buy some blue and some red pencils and altogether it costs me 90 pence.

If b is the number of blue pencils bought and if r is the number of red pencils bought, what can you write down about b and r?

$5b + 6r = 90$	10%
Two correct pairs, of	
(6, 10), (12, 5), (18, 0), (0, 15).	1%
$b + r = 90$	17%
$6b + 10r = 90$ or $12b + 5r = 90$	6%

minimum, as specific unknowns: 'I bought b blue pencils which therefore cost 5 × b pence altogether' etc. It was not possible to classify all the different responses that children gave, but of the answers that were coded, 'b + r = 90' and answers like '6b + 10r = 90' are probably the most interesting.

Most of the children who wrote 'b + r = 90' could cope with Level 3 items, yet their answer only seems to mean 'blue pencils and red pencils cost 90 pence' which though true, gives only limited information and uses the letters as objects. ('b + r = 90' *could* be read as 'the *number* of blue and red pencils bought cost 90 pence', but this is still very much tied to the concrete reality of the question, and is not a 'pure' statement about numbers. The number b + r does *not* equal the number 90).

Children who gave answers like '6b + 10r = 90' had found one correct pair of values for b and r (6, 10) but instead of expressing this in a form that showed that b and r are numbers (b = 6, r = 10) were essentially saying '6 blue pencils and 10 red pencils cost 90 pence' – again the letters were being used as objects.

Letter as specific unknown

The previous three categories all describe ways of *avoiding* generalised arithmetic, by not using the letters as genuine unknowns. The opposite is true of the present category – even though the idea of a *specific* unknown number is still a rather primitive notion.

The use of a letter as a specific unknown has already been discussed for Questions 9(iv) (p = 2n), 14 (r = 15), 5(iii) (e + f + g = 8 + g), 13(viii) (3a − b + a =), and 13(v) ((a − b) + b =). This usage is also required to answer 4(ii) and 4(iii) correctly, but not 4(i) (Table 8.6).

Table 8.6   Children's responses (14 year olds)

4(i) (Level 2)		4(ii) (Level 3)		4(iii) (Level 4)	
Add 4 onto n + 5		Add 4 onto 3n		Multiply n + 5 by 4	
n + 9	68%	3n + 4	36%	4n + 20 or 4(n + 5)	17%
				4n + 5 or 4 × n + 5	19%
		7n	31%	n + 20	31%
9	20%	7	16%	20	15%

It may seem surprising that 4(ii), in particular, turned out to be quite so difficult (Level 3). The answer 3n + 4 appears to be very simple, but for this very reason it is also rather unsatisfactory. In a sense nothing has been done with the 3n and the 4 to arrive at the answer, but since n is an unknown children have to recognise that this is all that *can* be done to combine the elements. Many children were unable (or willing) to do this, and instead gave the answer 7n or just 7 in which the elements that were meaningful (the numbers 3 and 4) were 'properly' combined but the letter was simply left as it was or ignored entirely.

These last answers can be said to belong to the category 'letter not used', and the same applies in 4(iii) to the answers n + 20 and 20. However, this approach is sufficient to answer 4(i) correctly (here it is legitimate just to combine the numbers, with the letter left as it is).

Though 4(ii) and 4(iii) both require the letter to be interpreted as a specific unknown, 4(iii) (Level 4) is significantly harder than 4(ii) because of its greater structural complexity. The operation x 4 has to be applied consciously to *both* elements of the expression n + 5. Instead, many children just 'attached' the operation to the expression as a whole, which corresponds very closely to all that is required to produce the correct answer 3n + 4 in 4(ii) but which in this case produced answers that were ambiguous, like 4 x n + 5.

It might be argued that answers like 4 x n + 5 simply arise through a lack of familiarity with the appropriate notation – in this case brackets. But this is difficult to sustain for children at the end of their 3rd year of secondary school, particularly as the ambiguity can be resolved without brackets. Rather it would seem that such children have never understood the significance of brackets (see for example Kieren, 1979).

Letter as generalised number

In contrast with a letter as specific unknown, where the letter is thought of as having a particular (but unknown) value, a letter used as a generalised number is able to take more than one value. (A distinction can be made between the idea of a letter taking on several values in turn and a letter representing a set of values simultaneously. However, this is not done here, although it seems to be the second idea rather than the first that later forms part of the concept of a variable.)

Questions 16 and 18(ii) (below) both seem to require the letters to be seen as generalised numbers. The two items were more difficult than many of the specific unknown items, and it *may* be the case that children get an understanding of specific unknown first. However, it is perhaps more fruitful to regard these two interpretations as 'different sides of the same coin' as it seems likely that in the course of many algebra problems children will flip from one interpretation to the

Table 8.7   Children's responses (14 year olds)

16 (Level 3)	%	18(ii) (Level 4)	%
What can you say about c if c + d = 10 and c is less than d?		L + M + N = L + P + N is Always Sometimes Never true (when)	
c < 5 c = 1, 2, 3, 4 (systematic list) c = 10 − d	11 19 4	Sometimes, when M = P	25
Unsystematic list	1	Sometimes. Or M and P given a specific value	14
One value only (usually c = 4)	39	Never	51

other, depending on which is momentarily the more convenient. (Thus, returning for a moment to Question 22, children might well find the answer $5b + 6r = 90$ by treating b and r as if they were specific unknowns, but at the same time realise that this answer covers *all* values of b and r.)

The point of Question 16 was to find whether children would find several values for c. As can be seen from Table 8.7, children most commonly found just a single value, although this is not to argue that some would not have been willing to find other values if asked.

18(ii) was adapted from an item of Collis. The item proved to be remarkably difficult, which Collis argues is due to the improbability, in children's eyes, of two distinct, unknown sets (M and P) having any values in common.

Letter as variable

The blanket use of the term 'variable' in generalised arithmetic is a common practice which has served to obscure both the meaning of the term itself and the very real differences in meaning that can be given to letters.

The concept of a variable clearly implies some kind of understanding of an unknown as its value changes, and if this is to go beyond the ideas already present in seeing a letter as a specific unknown and generalised number, it would seem reasonable to argue that the concept implies, in particular, some understanding of how the values of an unknown change, though precisely what this might mean is hard to pin down. One reason why the concept is so elusive is because many items that might be thought to involve variables can nonetheless be solved at a lower level of interpretation. Furthermore, it is often convenient to switch from one interpretation to another in the course of solving a problem, which makes it difficult for an observer, and for the individual himself, to disentangle the real meaning being used.

Question 22 suffers from precisely these difficulties. The statement about the numbers of blue and red pencils bought, $5b + 6r = 90$, can be derived by regarding the letters as specific unknowns or generalised numbers. However, neither interpretation brings out the full extent of the relationship that exists between b and r, for which it is necessary to take the interpretation of the letters several steps further.

With the letters regarded as specific unknowns, the relationship $5b + 6r = 90$ is seen simply as a statement which happens to be true for a particular, albeit unknown, pair of numbers. This statement is essentially static, it involves no idea of change. Alternatively, when the letters are used as generalised numbers, $5b + 6r = 90$ becomes a statement that is satisfied by several, but still essentially isolated, pairs of numbers, namely some or all of (6, 10), (12, 5), (0, 15), (18, 0). This view involves the idea that the values of b and r can change, but does not in itself indicate *how* they change, for which it is necessary to compare the values with one another in some way.

A first step in such a comparison might be to order the pairs of values, as in the diagram, from which it is possible to recognise a correspondence of the sort 'as b increases, r decreases'.

$$
\begin{array}{cc}
0 & 15 \\
\downarrow \;\leftrightarrow\; \uparrow & \\
6 & 10 \\
\downarrow \;\leftrightarrow\; \uparrow & \\
12 & 5 \\
\downarrow \;\leftrightarrow\; \uparrow & \\
18 & 0
\end{array}
$$

However, it is possible to go further still and describe the degree to which b and r change by establishing a relationship between b and r. Such a relationship might be 'the increase in b is greater than the (corresponding) decrease in r', or 'an increase in b of 6 is 1 more than the corresponding decrease in r of 5', or even 'the increase in b is 6/5 of the decrease in r' etc.

The added meaning that relationships of this kind give to $5b + 6r = 90$ is a genuine advance over interpreting the letters as specific unknowns or generalised numbers, and it was decided to regard the letters used in this way as variables. An important feature of these relationships is that their elements are themselves relationships, so they can be called '*second-order* relationships' (' "12 is greater than 6" by more than "5 is less than 10" '). This characteristic provides a useful operational definition of variables, in effect 'letters are used as variables when a second (or higher)- order relationship is established between them'.

Having found a way of defining the concept of a variable the problem remained that items involving variables could often be solved in a simpler way, and it proved very difficult to devise items where it could reasonably be assumed that variables were required to solve them. The best item in this respect was Question 3 (see Table 8.8), which was deliberately worded in such a way as to minimise this problem.

Table 8.8   Various responses to Question 3 (14 year olds)

Which is larger, 2n or n + 2? Explain.	
Correct, conditional response (eg 2n, when n > 2)	6%
2n	71%
n + 2 or 'the same'	16%

The point of this question was to see whether children would recognise that the relative size of the two expressions (2n and n + 2) was dependent on the value of n. As can be seen from the Table, most children wrote that 2n was the larger, and their explanation was usually of the sort 'because it's multiply' (which is really quite reasonable). Few of these children quoted specific values of n and there was little evidence of children having used any form of trial and error. In particular, it is very unlikely that children who solved the problem did so through a lucky guess (by hitting on n = 1 or 2) since most of the successful children also did extremely well on the rest of the test.

How then did children solve the item successfully and what made them hesitate and consider the effect of n instead of simply choosing one of the expressions as being the larger? The answer proposed here is that they were able, in effect, to establish a second-order relationship between 2n and n + 2.

The relevance of such a relationship can best be explained by seeing what happens to 2n and n + 2 when specific values are chosen for n. Consider, for example, the values n = 4 and n = 7 which give the pairs (8, 6) and (14, 9) for (2n, n + 2). Here the obvious (first-order) relationship, which holds for each pair in turn and which is prompted by the original question, is that 2n > n + 2. However, it is also possible to establish a second-order relationship between the pairs, which can be expressed as 'as n increases the difference between 2n and n + 2 increases (14 − 9 > 8 − 6)', or 'the increase in 2n is greater than the increase in n + 2 (14 − 8 > 9 − 6)'.

The significance of this relationship is that it opens up the possibility that for some smaller value of n there may be no difference between 2n and n + 2 (when n = 2), or the difference may even be reversed (when n < 2).

The argument here is not that children go through precisely these steps but rather that children who have sufficient 'processing capacity' to be able to cope with complex relationships of this sort are likely to consider the possible effect of n on the relative size of 2n and n + 2, whereas children without this capacity will go for something simpler and more immediate.

In summary, the point has been made that relationships of different order can be established between expressions like 2n and n + 2 or between the letters b and r in Question 22. The particular importance of second-order relationships is that they give an indication of the degree to which one set of values varies as a result of changes in another set, and it was therefore decided to define variables in terms of such relationships. The point has also been made that though it is common practice to call all letters 'variables', there are many generalised arithmetic tasks that can be solved without having to construct second-order relationships.

Levels of understanding

The test was subjected to a statistical analysis using some of the methods indicated in Chapter 1. Twenty-one of the 51 items in the test were rejected, either because they came from a question with a large number of similar items or because their correlations were relatively low. This left 30 items which were sorted into four groups. Each group covered a different facility range, with the cut-off points adjusted according to the complexity of the items and the nature of their elements. The groups can be regarded as representing different levels of understanding of generalised arithmetic, which are described below.

Level 1

The items at this level are shown in abbreviated form in the diagram below,

Table 8.9   Level 1 items

Item			Facilities (%)		
			13 yrs	14 yrs	15 yrs
8	$2\diagdown\overset{10}{\underset{9}{}}1$	$p =$	95	97	96
5(i)	$a + b = 43, a + b + 2 =$		92	97	95
9(i)	$\overset{e\diagup\diagdown e}{\underset{e}{\triangle}}$	$p =$	91	94	93
6(i)	$a + 5 = 8,$	$a =$	86	92	93
7(ii)	$6\;\boxed{}\;10$	$A =$	79	89	90
13(i)	$2a + 5a =$		77	86	87

together with the facilities for each year group tested. As can be seen, the items were extremely easy.

The items at this level are purely numerical (8 and 7(ii)), or they have a simple structure and can be solved by using the letters as objects (9(i) and 13(i)) by evaluating the letter (6(i)) or by not using the letters at all (5(i)). For more complex items children at this level tended to give answers like 4ht or even 5ht instead of 4h + t to item 9(ii), 8ab instead of 3a + 5b to 13(iv) and 763 instead of 761 to 5(ii). In the case of items that required specific unknowns these children were likely to evaluate the letter (p = 32, etc. instead of p = 2n in 9(iv), e + f + g = 12, etc. instead of 8 + g in 5(iii)) or not use the letter at all (7 or 7n instead of 3n + 4 in 4(ii)).

Level 2

The clear difference between these items and those at Level 1 is their increased complexity, though the letters still only have to be evaluated (11(i) and 11(ii), see below) or used as object (9(ii), 7(iii), 9(iii), 13(iv)). Children at this level could still not consistently cope with specific unknowns, generalised numbers or variables.

It might be argued that the advance made at this level is due simply to an increased familiarity with algebraic notation. However, the children who coped with the Level 2 items also performed more successfully on the test as a whole. More significantly, their average raw score on the Calvert DH IQ test, which can be regarded as an indicator of ability external to the algebra test, was also higher. This suggests that the use of correct syntax is, at least in part, conceptual.

The improvement in the Level 2 answers over those of Level 1 can also be seen as a first indication (which is much more fully realised at Level 3) of a willingness to accept answers which are to some extent 'incomplete' or ambiguous, which Collis describes as 'acceptance of lack of closure'. Thus, for example, many of the

Table 8.10   Level 2 items

Item		Facilities (%)		
		13 yrs	14 yrs	15 yrs
15(i)	(This was a numerical item concerned with diagonals of polygons)	63	75	72
9(ii)	[figure: house shape with sides labelled h, h, h, h and base t] p =	58	68	73
7(iii)	[figure: rectangle with side n and base m] A =	54	68	76
9(iii)	[figure: house shape with top sides u, u, sides 5, 5 and base 6] p =	54	64	67
11(ii)	m = 3n + 1, n = 4, m =	44	62	67
11(i)	u = v + 3, v = 1, u =	49	61	70
13(iv)	2a + 5b + a =	40	60	66

children who wrote 8ab (20 per cent) as a simplification of 3a + 5b + a, (of whom about three-quarters were at Level 1) may well have known how to write 3a + 5b but preferred their answer because it looked more 'proper' and complete. A similar argument applies to Question 11. In 11(i) (u = v + 3, v = 1, u =) most Level 1 children left the item out completely or gave the answer 2 instead of 4. These responses can both be explained by the ambiguity or 'openess' of u = v + 3 ('one unknown is 3 more than another unknown'), which may well have put many children off the item entirely, and which may have prompted others to read the equation as 'u and v together equals 3' which reduces the ambiguity but leads to the answer u = 2.

Level 3

The major advance made by children at this level is that they can use letters as specific unknowns, though only when the item-structure is simple. These children are able to regard answers like 8 + g, 3n + 4, p = 2n as meaningful, even though the letters represent numbers and not objects, and despite the lack of closure of the answers.

Table 8.11   Level 3 items

Item		Facilities (%) 13 yrs	14 yrs	15 yrs
15(ii)	A figure with k sides has . . . diagonals (extension of 15(i))	34	52	54
13(viii)	$3a - b + a =$	27	47	56
13(ii)	$2a + 5b =$	29	45	51
5(iii)	$e + f = 8$, $e + f + g =$	25	41	50
14	$r = s + t$, $r + s + t = 30$, $r =$	30	41	39
9(iv)	n-sided figure, each side of length 2. $p =$	24	38	41
4(ii)	Add 4 onto 3n	22	36	41
16	$c + d = 10$, $c < d$, $c =$	21	30	35

Level 4

At this level children can cope with items that require specific unknowns and which have a complex structure (13(v), 4(iii), 7(iv), see Table 8.12). They can also cope with items like 20, 22 and 17(i) which require, at a minimum, that the letters are regarded as specific unknowns, but where there is a strong temptation to treat them as objects (cakes and buns, pencils, wages, etc.). 18(ii) involves generalised numbers. In Question 21 it is necessary to realise that a set x can equally well be represented

Table 8.12   Level 4 items

Item		Facility (%) 13 yrs	14 yrs	15 yrs
18(ii)	$L + M + N = L + P + N$, Always, Sometimes (when), Never	11	25	27
13(v)	$(a - b) + b =$	15	23	32
20	What does $4c + 3b$ stand for (if cakes cost c pence each and buns b pence each, and if 4 cakes are bought, etc)?	14	22	30
4(iii)	Multiply $n + 5$ by 4	8	17	25
7(iv)	5 ▢ e 2 $A =$	7	12	16
21	If $(x + 1)^3 + x = 349$ when $x = 6$, what value of x makes $(5x + 1)^3 + 5x = 349$ true?	4	12	16
22	Blue pencils and red pencils . . .	2	11	13
3	Which is larger, 2n or $n + 2$? Explain.	4	6	10
17	(Question concerning a total wage, W, after h hours of overtime, given the basic wage and the rate for overtime.)	2	5	8

by an expression like 5x, and that, furthermore, this results in the transformation ÷ 5 (and not × 5) on the values or value of x. This shift in the usage of x, and the need to coordinate the operations involved, is in some ways similar to the coordinations required to use a letter as a variable which is tested in Question 3.

Summary

The items at Levels 1 and 2 can all be solved without having to operate on letters as unknowns, whereas at Levels 3 and 4 the letters have to be treated at least as specific unknowns and in some cases generalised numbers or variables. Within these pairs of levels the differences are structural. For example, whilst the letter in 'a + 5 = 8' (Level 1) can be evaluated immediately, in 'u = v + 3, v = 1' (Level 2) the child first has to cope with an ambiguous statement. Also, whilst 'add 4 onto 3n' (Level 3) only requires that a single operation is 'attached' to what is given, in 'multiply n + 5 by 4' (Level 4) this leads to an ambiguous answer (e.g. n + 5 × 4) and it becomes necessary to coordinate two operations.

Performance by year

Table 8.13(i)   Percentage of children at each algebra level

	13 yrs	14 yrs	15 yrs
Level 0	10	6	5
Level 1 (4/6 items)	50	35	30
Level 2 (5/7 items)	23	24	23
Level 3 (5/8 items)	15	29	31
Level 4 (6/9 items)	2	6	9

Table 8.13(ii)   Cumulative percentage of children at each algebra level

	13 yrs	14 yrs	15 yrs
Level 1 or above	89	93	93
Level 2 or above	40	58	63
Level 3 or above	17	34	40
Level 4	2	6	9

The above Tables show the percentages and cumulative percentages of children at each algebra level for the 13, 14 and 15 year olds. From Table 8.13(ii) in particular, it can be seen that there is a sharp improvement from the 13 to 14 year olds, however the change in levels (and in total score) from the 14 to 15 year olds is small and statistically not significant. This suggests that many 13 year olds may have suffered from a severe lack of familiarity with generalised arithmetic, but that once a minimum degree of familiarity was reached (the 14 year olds) performance was dependent more on cognitive development than on the specific experiences of algebra commonly met at school. This is supported by the observation that many 13-year-old children found the test a puzzling and uncomfortable experience in contrast to the confidence generally shown by the older children who seemed at least to think they know something about algebra. An examination of 1st and 2nd year secondary school text books (e.g. the SMP lettered series) also suggests that much of the early work in algebra is presented as an afterthought to other work, for example on number patterns where from the child's veiwpoint it serves no useful purpose.

Correlations with other tests

The algebra test had correlations ranging from 0.6 to above 0.7 with the other CSMS mathematics tests (using Pearson's r) and a correlation of about 0.7 with raw scores on the Calvert DH test of non-verbal reasoning. This indicates that children's scores on any one of these tests may account for as much as half (0.7^2) the variance in their scores on the algebra test, which is high when it is considered that this equals the variance due to the combined effect of all the other factors that might be operating (such as differences in teaching, in test administration and other forms of test-error).

Piagetian levels

It is difficult to establish a direct link between the algebra levels and Piaget's stages of cognitive development since his major study of adolescent thinking focused on science rather than mathematics tasks. However, from general descriptions of these stages (that, for example, the child at the formal operational level can work with second-order operations and can think in possibilities rather than being tied to concrete reality), from Collis's research (which was within a Piagetian framework) and from empirical evidence (the comparison of performance on the algebra test and one of the Piagetian class tasks developed by the science wing of CSMS) it can be argued that the algebra levels correspond to the Piagetian sub-stages listed here:

Level 1   Below late concrete
Level 2   Late-concrete
Level 3   Early-formal
Level 4   Late-formal

The value of trying to establish such a correspondence is that it puts the analysis of children's understanding into a more general framework which might apply to other areas of mathematics and to areas outside mathematics. Also the framework might be familiar to the reader. At the same time, even if the framework is accepted as useful (see for example Halford, 1978), the algebra levels should not be regarded as some 'pure' measures of cognitive development, although the evidence presented in the previous two sections does suggest that, given a minimal degree of familiarity, the test provides a reasonable guide to children's cognitive levels.

This evidence also suggests that performance may be affected only to a small extent by a year of schooling (14 to 15 year olds), which raises the question of what experiences, if any, might bring about a more substantial change.

Implications for teaching

It is hoped that this research will help teachers see more clearly the diverse conceptual demands of seemingly commonplace activities in school algebra. The research has identified a number of different meanings that can be given to the letters in generalised arithmetic, the choice of which may depend to a large degree on children's cognitive levels. More generally, in algebra and in the other topics investigated, the research has found that children frequently tackle mathematics problems with methods that have little or nothing to do with what has been taught. This may be because mathematics teaching is often seen as an initiation into rules and procedures which, though very powerful (and therefore attractive to teachers), are often seen by children as meaningless. It follows that children's methods and their levels of understanding need to be taken far more into account, however difficult this may be in practice.

On the algebra test the majority of 13, 14 and 15 year olds were at Levels 1 and 2 (73, 59 and 53 per cent respectively) and were not able to cope consistently with items that can properly be called algebra at all, i.e. items where the use of letters as unknown numbers cannot be avoided. In Piagetian terms these children would seem to be at the stage of concrete operations, which means that for most children the teaching should be firmly rooted in this level whether the aim is to consolidate their understanding or to ease the transition to formal operational thought. An example of the kind of activity that this implies (which is currently being investigated by SMP) is shown in the diagram below.

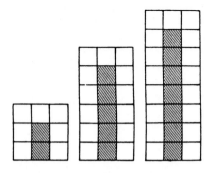

Here children are presented with patterns of, say, black and white tiles which conform to a rule, and are asked to find the number of white tiles needed for perhaps 10, 20 and eventually 100 black tiles. The strength of such a problem is that the configuration are easily defined and recognised (as 'bridges' of some sort perhaps) and easily constructed (when the numbers are small), so that a firm basis exists for gradually extending the work to a point where a practical construction becomes both tedious and unreliable. At the same time the numerical relationship between the tiles is far from obvious and can be derived in different ways (add $1\frac{1}{2}$, then double or double, then add 3) so that the task of finding the relationship, of representing it economically and unambiguously, and of comparing the equivalent representations provides children with a worthwhile challenge.

It also seems sensible to base the teaching given to children at Levels 1 and 2 on the meanings for the letters that these children readily understand. On closer examination this is by no means a straightforward task, for example, the use of letters as objects totally conflicts with the eventual aim of using letters to represent numbers of objects. However, it may well turn out to be the case (see Inhelder et al, 1974) that it is precisely through being made aware of such conflicts that children see the need to reorganise their thinking and thereby move towards a higher level.

9   Graphs

Introduction

The interpretation and use of graphs is an aspect of mathematics which has appeared in secondary school syllabuses for many years. The use of coordinates is commonly ascribed to Rene Descartes (1596–1650), although it seems that the notion of a coordinate system played little, if any, part in his work. It is known that a rectangular coordinate system was used by Hipparchus (c 161–126 B.C.) to locate places on the earth's surface, using a 'longitude', the distance from east to west along the Mediterranean, which was the 'length' of the then known world, and his place of writing, Rhodes, as a meridian. It was probably Professor G. Chrystal in his *Algebra: An Elementary Text-book for the Higher Classes of Secondary Schools and Colleges* published in 1886 who first placed the study of graphs in the school syllabus in this country. He introduced 'a new way of looking at analytical functions, the graphical method as it is called'.

The items for the test 'Graphs' were selected in order to investigate the important underlying ideas which are necessary components of the understanding of graphs in schools today. The questions used involved ideas of coordinates, the use of scales, the idea of rate and gradient, continuity and the use of algebra and equations. These ideas were investigated at various levels of difficulty, but were often presented in unfamiliar forms, so that the pupils could not rely always on learned techniques.

The questions

After the usual series of individual interviews and preliminary class tests, the paper was eventually given to 459 second year pupils (13 yrs old) 755 third year pupils (14 yrs old) and 584 fourth year pupils (15 yrs old) from a number of different secondary schools. The facilities quoted below are based on their results.

Block graphs and coordinates

Very little difficulty was experienced by the pupils with elementary items on block graphs, and the use of rectangular coordinates to plot points, when the numbers involved were integers. About 90 per cent of the pupils were successful at such items.

The test also included an item on non-rectangular coordinates, and an unfamiliar example, which the pupils were unlikely to have met before, was chosen in order to see whether they could apply what they already knew about coordinates. They were given the coordinates of the point A (4,6) in the diagram below and then asked for the coordinates of point B. They were also asked to plot the point (7,6).

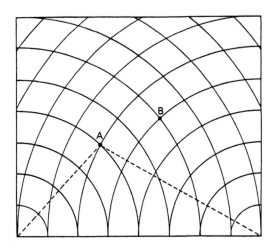

Percentage successful

13 yrs	14 yrs	15 yrs
19.5	31.0	38.0

Clearly many pupils found it difficult to adapt what they knew to this strange situation.

Continuous graphs

The transition from the idea of drawing a graph of discrete quantities to that of a continuous quantity is an important feature of secondary mathematics courses. There were several items in the test which attempted to investigate the pupils' understanding of continuity. One such item was Question 3 (see below). The intention was to see whether, when the pupils had plotted points and joined the line on which they lay, they were able to recognise that other points also lie on the line and that there are indeed infinitely many points on a line and between any two points on the line.

3. (Diagram drawn on a 2 mm grid.)
3.1 Plot the points (2,5), (3,7), (5,11).
3.2 These points lie on a straight line. Draw the line. Find some other points on the line and write them down.
3.3 The point (4.6,10.2) also lies on the line. Mark its position approximately.

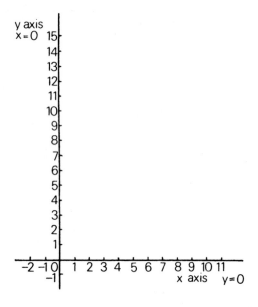

3.4 Plot the point $(1\frac{1}{2},4)$.

3.5 How many points do you think lie on the line altogether?

3.6 Are there any points on the line between the points $(2,5)$ and $(3,7)$? If so, how many?

Facilities (percentage)

	13 yrs	14 yrs	15 yrs
3.1	94.9	91.2	89.1
3.3	36.0	27.9	34.6
3.4	79.2	76.9	80.1
3.5	6.4	6.2	19.6
3.6	4.1	4.0	11.1

Preliminary interviews suggested that many pupils found difficulty with the idea that there are any more points on the line other than those they had plotted. Several said that there were no points between $(2,5)$ and $(3,7)$, while others thought there was just one, presumably the mid-point. For the number of points altogether, several gave the actual number of points they had plotted, while others gave the number of 'integral' points on the page and some counted the number of points where the line crossed the grid. In order to make it clear to the pupils that points other than those with integral coordinates were to be considered they were asked, in later drafts of the test, to plot the points $(1\frac{1}{2},4)$ and $(4.6,10.2)$, being told that both also lie on the line. It is worth noting that they found the decimal coordinates much more difficult than the fractional ones. Many interpreted the point $(4.6,10.2)$ as two separate points $(4,6)$ and $(10,2)$, neither of which, of course, lie on the line! It seems unlikely that most of the pupils had not met decimals, more probably they were not used to them in this context, and adopted the common strategy of making something unfamiliar into something familiar, even though wrong! Interview responses to Question 3.5 included a range from those who said, as indicated above, some small finite number, to those who said 'lots' or 'hundreds', and then those who appear to have some idea of infinity by using the word or an expression like 'you could go on counting for ever'. Philip (age 12) said, for 3.5, '18, if you just take the whole numbers, but more if you can go into fractions and extend the graph', and to 3.6 'Too many to count if you count fractions'. David (age 14) said to 3.5 'I think there can be as many as possible' and to 3.6 'I think there is one main one', while Anne (age 14) said to 3.5 '15½ points lie on the line'! Even those who are aware of the existence of very many points appear to be limited in their idea of how many by the physical constraints of actually drawing the points, for example Susan (age 12) said of 3.5 'As many as you can put there' and of 3.6 'As many as there is room for'. The percentage of pupils giving the various types of answer is of interest. They are, for Question 3.5:

Table 9.1   Pupils' understanding of infinity (i) (percentage)

Response	Some finite number e.g. 4, 5, 8 . . .	'Hundreds', 'lots', etc.	'Infinite' − 'As many as you like' etc.
13 yrs	63.8	12.6	6.4
14 yrs	69.2	5.1	6.2
15 yrs	51.6	9.7	19.6

and for Question 3.6, which concerned the number of points on the line between two given points:

Table 9.2   Pupils' understanding of infinity (ii)(percentage)

Response	'None'	'1'	Some finite number	'Infinite' − 'As many as you like' etc.
13 yrs	15.4	36.8	27.5	4.1
14 yrs	17.7	41.8	22.1	4.0
15 yrs	13.1	34.4	21.3	11.1

Scattergrams

Another item used the idea of a scattergram − which not many school text books use − in order further to investigate the pupils understanding of continuity. Here they were asked whether the points should be joined, and whether one could interpret points between the ones plotted.

5.   Ann drew a diagram to show the height and waist measurement of herself (A), Brian (B) Cheryl (C), David (D) and Frank (F). (Diagram drawn on 1 cm grid.)

5.1   What is Brian's height?

5.2   What is Brian's waist?

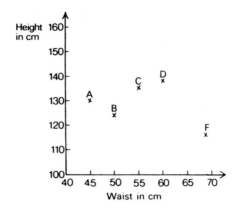

5.3   Mark in George (G) whose height is 150 cm and waist is 70 cm
5.4   What can you say about Frank's appearance?
5.5   Should we join up the points on the diagram?
5.6   Why do you think this?
5.7   What can you say about the height of a child whose waist is 65 cm?

Facilities (percentage)

	13 yrs	14 yrs	15 yrs
5.1	61.2	63.6	73.6
5.2	93.3	86.7	91.3
5.3	92.0	89.4	95.2
5.4	82.5	77.6	88.1
5.5 } 5.6	54.8	46.1	54.2
5.7	11.6	10.3	19.9

It will be seen that most pupils were able to read the information from the graph correctly, but the answers given to 5.5/5.6 were very interesting. Some of the replies given during the interviews are given below, for some children the appearance of the graph seemed to be of paramount importance. In reply to the question 'Should we join up the points?' Philip (age 14) answered 'Yes, I think this because it would be more accurate and look neater', Jane (age 13) said 'Yes, because it would be tidier', but Simon (age 13) answered 'No, because the points would be all in a mess'. Others clearly thought that points are only joined if they lie in a straight line or make some other recognisable pattern like Sharon (age 14) 'No, because they don't all go in a straight line', Stuart (age 13) 'No, because they don't make any particular shape' and Alan (age 14) 'No, the points are all over the place'. On the other hand, some, like Josephine (age 13), said 'Yes, because it is something you always do'. Many, though, clearly understood the nature of a scattergram, e.g. Philip (age 15) 'No, they are separate people', Jillian (age 14) 'No, they are different people' and Alison (age 15) 'No, because everyone's different, no-one's connected'. The responses to 5.7 varied from giving an actual height to the child, through saying 'he is quite tall', 'he could be quite tall' to 'you can't say' or 'nothing'. It was thought that some pupils might even talk about 'lines of best fit' or regression lines, but there was no instance of this in a recognisable form.

The percentage of pupils who thought the points should be joined up, for some reason or other were:

13 yrs	14 yrs	15 yrs
21.6	29.0	23.7

The commonest reason was based on the appearance of the graph.

Choice of axes and scales

Most C.S.E. and many 'O' level questions which require the candidate to draw a graph give their candidates the scale to be used. While this is undoubtedly a help to the candidates, it must surely also ease the task of the markers! It eliminates the very awkward scales which some children choose and which sometimes make it impossible to read the information required.

One item in the test required the pupils to plot the points (20,15), (−14,3) and (5,−12) on a sheet of graph paper which was supplied. The purpose of the question was to find out how able the pupils were to select their own scale, and to site their own axes so that the negative numbers as well as the positive numbers could be plotted. Suitable scales were taken to be either 5 to 1 or 10 to 1, and scales such as 3 to 1, 4 to 1 and 7 to 1 were regarded as unsuitable. The percentage of pupils giving suitable and unsuitable scales is given below, together with the percentage who produced suitably placed axes.

Table 9.3   Scale selection (percentage)

	Suitable scale	Unsuitable scale	Correct axes
13 yrs	32.9	10.8	27.2
14 yrs	49.8	7.9	35.7
15 yrs	61.3	13.3	65.4

Some thought it was possible to have different scales on the positive and negative parts of the axes, pupils interviewed produced, for example:

Julie (age 13)

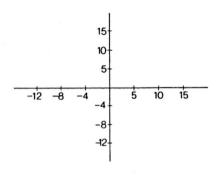

Carlton (age 14)

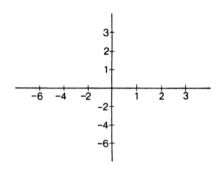

David (age 13)

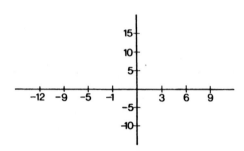

Adrian (age 14)

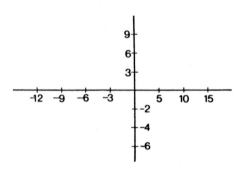

Some failed to appreciate that the scale should start at the origin, and began with the first number on their scale, such as

Anthony (age 14)

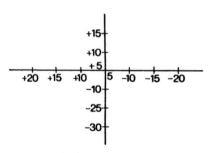

Some failed to realise that negative numbers were required and did not allow space for them, while others overcame the problem by ingenious means:

Carol (age 13)

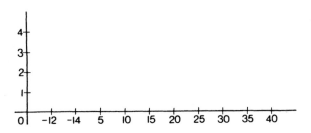

Change of axes and change of scale

Two other items in the test were aimed at finding out whether pupils realised the effect of changing the axes or changing the scale of a graph. The scattergram mentioned earlier was redrawn with the axes reversed, and the pupils were asked to describe the appearance of a person whose height and weight were plotted. This was exactly comparable to a question with the axes in their original position. The facilities in the two cases were:

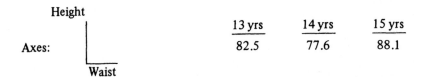

	13 yrs	14 yrs	15 yrs
Axes: (Height / Waist)	82.5	77.6	88.1

	13 yrs	14 yrs	15 yrs
Axes: (Waist / Height)	59.4	55.6	69.0

However there is evidence that many pupils were misled by the visual appearance of the graph, heights usually being measured 'up' the page. This will be referred to in the next section.

In another item the pupils were given the following three straight line graphs and asked which were the two that showed the same information:

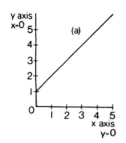 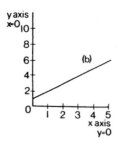 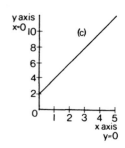

Here, of course, the first and last look similar, but the scale on the y axis has been changed. The percentage of pupils who realised that the first two, in fact, gave the same information was:

13 yrs	14 yrs	15 yrs
46.4	63.4	68.5

Distance-Time graphs

Travel graphs form part of most courses on graphs. One of the questions was of a fairly common type in which there was a description of a journey together with its graph and pupils were asked to pick out different rates of travel, arrival times and so on. The majority of pupils managed this quite well.

However, the earlier interviews had suggested that several pupils found difficulty with travel graphs, even though they appeared to be able to give the correct answer. It was clear that several of those interviewed had incorrect perceptual interpretations of the graph. Some thought of the graph as a journey that was up and down hill, or as directional on the ground, and found it difficult to deal with the abstract notion of 'distance from an origin', in this case, the boy's home.

In order to pursue the idea of distance-time graphs further, therefore, it was decided to give them some graphs which could not represent journeys, and ask them to say what they thought they showed. The graphs used were:

7.   Which of the graphs below represent journeys? Describe what happens in each case.

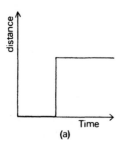 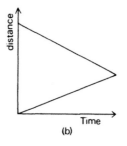 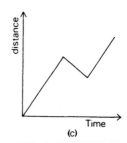

Facilities (percentage)

	13 yrs	14 yrs	15 yrs
(a)	9.5	8.4	15.0
(b)	11.1	9.3	15.7
(c)	14.7	17.2	25.2

In 7(a), pupils were expected to say that the graph did not represent a journey, and a few did so. Some also added that a certain distance appeared to have been travelled in zero time. However, many interesting misconceptions emerged from the answers of those who did not get the item correct. The 'journey' was often described as 'going east, then due north, then east' or 'went along, then turned left, then turned right' or 'went along a corridor, then up in a lift, then along another corridor'. For 7(b) there were similar replies, reference often being made to 'going along, then turning left' or 'going North-east, then North-west', or 'going back the way he came', while 7(c) produced 'climbing a mountain' or 'going up, going down, then up again' and so on. The number of children making such a spatial response suggests that many, when confronted by a travel graph, have incorrect visual impressions of what is being represented. The percentage of such pupils giving such incorrect spatial interpretations is given in the Table below.

Table 9.4   Incorrect spatial responses to distance/time graph

	13 yrs	14 yrs	15 yrs
7(a)	16.2	16.2	12.1
7(b)	9.2	6.7	9.6
7(c)	12.6	14.3	11.4

There is some evidence that those children who are particularly strong visualisers — that is they seem to think in visual terms — found extra difficulty with this question and with some other items in the test where a graph can be visually misleading. Another example is the item in the previous section where the axes and scale of a graph were altered and the appearance therefore changed.

Gradient

Some of the items in the test concerned the understanding of the idea of gradient. One concerned the growth of a plant in which the pupils had to decide which part of the graph represented the fastest rate of growth, and to complete a constant rate graph.

Another question, illustrated overleaf, investigated whether the pupils realised that the gradient of a straight line was the same at all points on the line, and whether they could use the idea of gradient to decide whether two lines were parallel. The question was:

10. Here is the graph of y = 3x − 5.

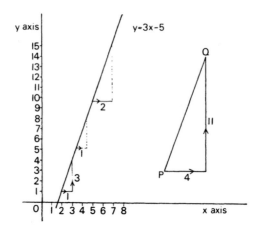

10.1 How many units is the line of dots on the graph?
10.2 How many units is the line of dashes? [Upper section.]
10.3 Is the line joining PQ parallel to the first line?
10.4 Why do you think this?

Facilities (percentage)

	13 yrs	14 yrs	15 yrs
10.1	28.8	34.7	36.1
10.2	26.5	31.5	36.8
10.3 } 10.4 }	3.1	3.2	9.2

During the interviews, most children found the lengths of the uprights by actually measuring on the graph paper and only very few realised that the second 'dashed' line could be found by doubling the 'dotted' line. When prompted some agreed that they could have doubled. When marking 10.1 and 10.2 on the test it was not possible to tell whether a child had measured or calculated if he gave the answers '3' or '6' respectively. But answers such as 2.9 or 6.1 were counted as incorrect since they, presumably, had been measured. It can be seen that only about a third of the pupils did give the answers '3' or '6', and so it is perhaps not surprising that only a very small percentage were able to give a correct response to 10.3/10.4, mentioning the 1:3 and 4:11 ratios in some way or another. The lines were deliberately drawn so that they looked superficially parallel, but the numbers involved were made simple. It was thought that most children would know the (1,3), (4,12) relationship.

Parallel lines

Many interesting ideas of parallel lines emerged both during the interviews and in the tests. Several children thought the lines in Question 10 could not be parallel as one was longer than the other, while other replies are exemplified by Andrew (age 13) 'No, because the line is diagonal', Jackie (age 14) 'No, because it is not straight', Nicky (age 14) 'No, because the first line is straighter than the second one' and even Lorraine (age 14) 'No, because if it was, the bottom and the top would be opposite'. Those who thought the lines were parallel for various incorrect reasons included Julie (age 12) 'Yes, because they both go together', Jo (age 13) 'Yes, because the lines have 3 marks on the y axis all the way along' — who both used the grid, but not the idea of gradient — but also Helen (age 12) 'Yes, because if it was bendy it wouldn't be parallel'. Some measured across the page, inaccurately and thought they were parallel, but only a few noticed the (1,3), (4,11) relationship and so came to the correct conclusion.

Equations of straight lines

Several questions in the test involved the use of algebra to describe a straight line, and all these items produced very low facilities. For example, the pupils were given the graphs $y = 2x$, $y = 2$, $x = 2$ and $x + y = 2$. They were asked to indicate which of the four was $y = 2x$, and to give the equations of the other three. The facilities were:

	13 yrs	14 yrs	15 yrs
$y = 2x$	18.8	18.2	26.9
$y = 2$	16.7	10.3	14.3
$x = 2$	15.2	9.8	14.0
$x + y = 2$	5.9	3.9	5.1

The success rate on these items can be seen to be very low. Rather more managed to attach $y = 2x$ to the first line than to identify the rest, but of course this was the only one for which an equation was provided. The preliminary interviews suggested that the success rate on this item would be low. Many had no idea what was intended, while others had very vague ideas about graphs. Philip (age 15) gave as his 4 equations, in order, $y = x$, $x = 2y$, $y = 2x$, $2x = 2y$, and clearly had some notion about graphs but was unable to attach the right equation to any of the lines and presumably thought $y = x$ and $2x = 2y$ were different lines. Jillian (age 13) gave $y = 0x$, $2y = x$, $y = 2x$ and $2y = 2y$, and explained that she looked at the values on the axes (the intercepts) and put these in front of the x's and y's (made them the coefficients) — her 'method' lacked consistency and was adapted to her expectation of what an equation should look like. Rebecca (age 15) gave $2y$ and $2x$ as the equations of the middle lines and others made no attempt at all. In the full test, about a third of the pupils made some attempt at writing an equation, although incorrect, leaving a substantial number who made no response to the question at all.

In another question the graphs were attached to actual problems to see whether this was easier. The problem given was:

A boy's wages, £W, in a supermarket are made up as follows: basic wage £3, plus £2 an hour for every extra hour, h, that he works. So W = 3 + 2h, which can be written as W = 2h + 3.
When h = 1, what is W?
When h = 2, what is W?

The children were then asked to plot the information on the grid provided and draw the graph of W = 2h + 3. The axes and scale were marked on the given grid. The facilities were:

	13 yrs	14 yrs	15 yrs
W, when h = 1	43.7	36.5	60.3
W, when h = 2	41.1	35.4	61.3
Points plotted	19.8	24.4	35.8
Line drawn	7.5	12.6	19.9

There were, then, a number of pupils who, although they had calculated the information, were not able to plot the points on the paper in this particular instance although, as mentioned earlier, they were in general well able to plot points when they were given as an ordered pair.

Immediately after this question the pupils were asked to plot the graph of y = 2x + 3, the grid with axes and scale being given as before. The facilities this time were:

	13 yrs	14 yrs	15 yrs
Points plotted	17.7	17.5	23.0
Line drawn	10.5	12.0	16.2

It thus appeared to make little difference whether they were presented with an 'abstract' graph, with the equation only being given, or whether it was wrapped up in a more 'concrete' situation.

Finally a question was included in an attempt to analyse whether pupils understood the procedure to find the solution of two simultaneous linear equations by a graphical method. The question was:

16.   Here are the graphs of y = x − 1 and x + y = 8

16.1   Find an x and y that make y = x − 1 true.

16.2   How many pairs of values of x and y are there that make y = x − 1 true?

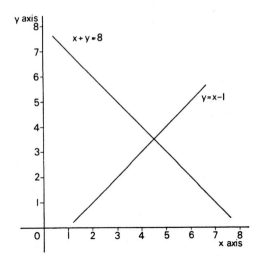

16.3 Find an x and y that make x + y = 8 true.

16.4 Find an x and y that make both y = x − 1 and x + y = 8 true at the same time.

16.5 How many pairs of values of x and y are there that make both y = x − 1 and x + y = 8 true at the same time?

Facilities			
	13 yrs	14 yrs	15 yrs
16.1	15.7	22.6	35.1
16.2	1.8	4.7	9.7
16.3	15.7	23.9	36.8
16.4	9.0	16.5	24.9
16.5	8.2	16.7	28.3

The question picks up the idea again, that there are an infinite number of points on a line but only one point of intersection. The children were asked to find values of x and y on the lines separately first in order to help them see the many possibilities. However, the results show that the solution of equations by this method appears to be a sophisticated idea, one which hardly anyone in this age range appeared to have. It highlights the difference between the relative ease with which children can be taught a process and to obtain the correct answer and the difficulty in getting them to understand the underlying concepts.

Levels of understanding

From all the items in the graphs test three groups of items have been identified using the method described in Chapter 1. Each group of items is called a level, and a child is assigned to the highest level in which it is successful on about two-thirds of the items.

Level 1 includes items 3.1, 3.4, which involve plotting points, others on interpreting block graphs, recognition that a straight line represents a constant rate and simple interpretation of the scattergram.

Level 2 includes simple interpolation from a graph, recognition of the connection between rate of growth and gradient, use of scales shown on a graph, interpretation of simple travel graphs and awareness of the effect of changing the scale of a graph.

Level 3 consists of items that require an understanding of the relation between a graph and its algebraic expression. This includes the items mentioned in the last section on the equations $y = 2x$, $y = 2$, $x = 2$ and $x + y = 2$ and on the graphs of $W = 2h + 3$ and $y = 2x + 3$.

The results were:

Table 9.5   Levels on graphs

	Facility range (percentage)	Pass mark
Level 1	78.5–97.6	5/7
Level 2	58.2–70.9	4/6
Level 3	12.8–26.5	7/11

98.7 per cent of the children at any level also succeeded on all lower levels. Level 0 was included for those who succeed at less than 5/7 of the items in Level 1 and who thus were unable to make any coherent attempt at the questions.

The percentage of children who achieved each level for the three year groups in 1976 was:

Table 9.6   Levels by year group (percentage)

	13 yrs	14 yrs	15 yrs
Level 0	6.5	9.0	4.0
Level 1	33.0	32.5	22.5
Level 2	55.0	48.5	57.0
Level 3	5.5	10.0	16.5

The items involving the idea of continuity and infinity were more difficult than Level 3, but since some pupils could answer these without being successful at Level 3 these items do not occur in the scale.

Comparison of performance by year groups

The fourth year had more pupils at Level 3 than the two younger age groups and fewer at level 0. The 15 year olds were the most successful at almost all the items. In those few items (6 out of 50) in which they did not score highest they were only marginally lower. The easiest items were almost equally well done by all age groups. There is also little difference between the performance of the 13 and 14 year olds in the most difficult items, with the 13 year age group sometimes being more successful than the 14 year olds. However, in these difficult items, there is a considerable increase in facilities for those aged 15. In one question involving the notion of infinity the fourth years had a facility of 13 per cent more than the 2nd or 3rd years, and they also did notably better on the distance-time graphs. Their facilities on the item on calculating W given h, for the relation $W = 2h + 3$, were also well up on those of the other years — some 25 per cent better. There were, however, only about 11 out of the 50 items in which the 15 year olds had a facility more than 8 per cent above those of the two younger years. Details of the results of the longitudinal study appear in Chaper 12.

General implications for the teaching of graphs

Graphs in the general curriculum

The evidence of the graphs test suggests that there are many aspects of graphs which are well within the capability of almost all secondary pupils. Certainly this is true of block graphs and other pictorial representations of numerical information and the plotting of points on a coordinate grid. However, there appears to be a large gap between the relatively simple reading of information from a graph and the appreciation of an algebraic relationship. Since several other subject areas, such as geography, environmental studies and science for example, often make use of graphs to deploy information, it is important that this distinction is understood. While many children will be able to read information from a graph or to plot given data, it seems that only a few will be able to understand the connection between an equation and a graph.

Visual appearance of graphs

It should be remembered that while the main purpose of a graph must be to illuminate numerical data by using a visual form, there are some occasions when a graph can be visually misleading. An example of this has been seen in the distance/time graphs which looked to many children, like journeys in a vertical plane. Another common problem is the fact that a distance/time graph for a stone thrown vertically in the air looks like this:

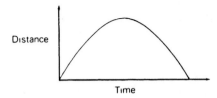

which is visually more like that of a ball being thrown from one person to another. It is important to ensure that pupils are not being actually misled by the graph.

Order of ideas

It is important that one should look carefully at the order in which graphical ideas are presented and try to relate them to the stage of development of the pupils. If one examines many of the commonly used textbooks one finds considerable differences. For example, one can find well-known texts which introduce polar-coordinates in each of the years 1, 2, 3 and 5 of a five-year school course. More seriously, some of the ideas shown in the graphs test which seem to be understood by only a very small minority of 4th year pupils appear much earlier than this in many books. Equations of straight lines are one such instance, in that they appear in the first year of one much used series and in year 2 of several others, yet the results show this to be too difficult for most pupils. If the same work came rather later in the course there might be a greater chance of it being understood.

It is possible, taking the levels of difficulty implied by the results of the graphs test, to suggest a syllabus for use in secondary schools covering the years 2, 3 and 4. If one considers the school population divided into those who will not be taking any external examination in mathematics (say the lower 30 per cent), those who will take an examination such as CSE (say the middle 50 per cent) and those who will take 'O' level in mathematics which includes those who will continue with mathematics, say an upper 20 per cent, then it would seem reasonable to expect each group to be able to understand the following:

Lowest 30 per cent − The reading and presentation of information in the form of a block graph, the use of coordinates, ideas of gradient and rate of growth, simple distance-time graphs, interpretation of scattergrams, reading information from a straight line graph including interpolation, simple scales and the effect of changing scale and axes

Middle 50 per cent − As above plus the reading of information from non-linear graphs, simple ideas of continuity, coordinate systems other than rectangular, interpolation from non-linear graphs

Top 20 per cent − As above plus continuity, knowing when points should and should not be joined by a curve, equations of straight lines, distance-time graphs, drawing graphs of straight lines, solution of linear simultaneous equations

Conclusion

One could argue that every child should know sufficient about graphs to enable him or her to appreciate the value of a visual display of information and be able to interpret such information when it appears in newspapers and magazines. For some this will be only a beginning, and will lead them on to a study of functions where they will find how powerful a tool graphical methods can be. A.W. Siddons (1933) said that the aim of teaching graphs should be 'to give the idea of the inter-dependence of two related quantities, to give ideas of continuity, and to give the power to deal with questions that are impossible or too difficult by ordinary methods'. This still remains an excellent aim for our abler pupils, while for the rest we might have 'to give the ability to present information which has been gathered in a form easily taken in by the eye, to give the power to absorb information presented in a visual way and not to be confused by misleading diagrams, and to develop an appreciation of the use of a coordinate system to locate a point in the plane in a variety of instances'.

10   Reflections and rotations

Introduction

Transformation geometry may not be the first topic that springs to mind at the mention of 'modern mathematics', but the reasons for its introduction and the ways in which it has been adopted provide a perfect illustration of the hopes and disappointments of those who have attempted to revitalise the content and teaching methods of secondary school mathematics over the last 20 years.

The CSMS research into children's understanding of transformation geometry (which included a questionnaire on the ways in which the topic is taught in schools) confirms the view that the acceptance of the topic, though widespread, has occurred with a lack of conviction in many schools and a reluctance to abandon traditional, expository methods of teaching (for example, only about half the schools in the survey who studied reflection seemed to have used paper-folding or mirrors). These tendencies might in part stem from the belief that geometrical transformations, once their meaning has been explained, provide little challenge to secondary school pupils. However the research evidence shows that this belief is mistaken. A more valid reason, particularly for the relatively low esteem of the subject matter, is that the study of transformation geometry was introduced not as a topic in its own right but as a means to a number of ends, none of which now seem very convincing.

The first of these ends was essentially negative. It was felt that the traditional, 'euclidean' geometry was not appropriate for the majority of secondary school children. This is not in dispute, though the criticism applies not to the subject matter itself but to the fact that it was being taught as a tight deductive system which most children could master only by rote learning. In the circumstances it is not surprising that there was resistance to the introduction of quite different subject matter (as well as less formal methods), particularly as the material being discarded formed on elegant mathematical system which was rightly seen as an important part of our cultural heritage.

A second reason for introducing the topic was the hope that children would discover general rules about the combinations of transformations which would provide insights into mathematical structure, particularly the structure of the group. However, most modern syllabi, at least up to O-level, do not demand that this should be fully realised, with the result that teachers and their pupils are left working towards goals that are confused and incomplete. Moreover, many children's difficulties with single transformations pre-empts them from even considering combinations unless very simple transformations are used.

A related reason for introducing the topic was the belief that transformation geometry would provide a coherent embodiment of matrix algebra and that this would be recognised as an impressive demonstration of the unity of mathematics. The link with matrices is impressive, and even magical. However, it is extremely doubtful whether it is seen as meaningful by the majority of secondary school children.

It would seem, therefore, that in introducing transformation geometry into the secondary school curriculum a major concern has been to ensure that the loss of one area of high status knowledge (deductive euclidean geometry), which was recognised as unsuitable, was made good by the introduction of other areas of high status knowledge – even though these are equally unsuitable. This concern to protect the status of school mathematics reflects the professional interest of mathematicians who, particularly through the examination system, still exert a strong control over curriculum reform. Unfortunately these interests do not necessarily coincide with those of the children who study the subject.

Two conclusions follow: either transformation geometry is accepted as relatively unimportant, since the aims for which it was introduced cannot be fulfilled, or the transformations are studied *singly* and are seen as valuable in their own right. It is hoped that this chapter will show that the latter is not only feasible but would provide a demanding course of study for most secondary school children.

The content of the test

The CSMS transformation geometry test was restricted to an investigation of reflection and rotation. Information on two other transformations (translation and enlargement) is available from the CSMS vectors and ratio tests.

The two main sections of the transformation geometry test were concerned with single transformations and both these sections were preceded by a number of practice items to which the answers were given. The third section consisted of questions involving combinations of reflections and rotations, some of which also required the use of inverses.

Many of the questions involved drawing, and here children were usually told just to sketch their answers. This reduced the reliability of certain items but it also meant that far more items could be included in the test, which was vital if the many factors influencing performance were to be identified. The instruction to sketch the answers was particularly relevant to the younger children, many of whom, when given the choice, elected to draw very neat answers which prevented them from completing the test. Also, their conceptual errors often meant that this care and effort was wasted.

For marking children's drawings, regions had to be defined to delimit what was to be regarded as correct. This was done by devising rules that took into account such factors as distance from the mirror-line, the slope of the mirror-line and whether or not a ruler was to be used. The resulting regions (an example of which is shown in the diagram opposite) were printed onto acetate sheets which could be placed over the pupil's scripts. The existence of these rules and the use of the acetate means that the assessment of pupils' responses can be regarded as 'objective'. However, clearly quite different rules could have been devised, and this needs to be borne in mind when the percentage of pupils deemed to have answered an item correctly is considered.

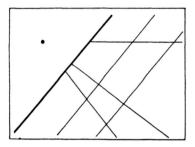

In addition to defining these regions, a marking scheme was devised that was sensitive to the different *ways* in which pupils arrived at their answers and to the different *kinds* of answers that they produced. It took many drafts and a thorough examination of several hundred scripts before a satisfactory marking scheme emerged, and here again it is important to realise that a quite different scheme, yet one appearing to work equally well, could have been devised.

The diagram below gives an indication of the scheme that was used for one of the reflection items (A1. 8), and also shows the percentage of 14 year olds giving each type of response. (For convenience, the results quoted throughout this chapter will be for this age group, unless it is stated otherwise.)

code1	code2	code3	code4	code5	code6	code7	code8	code9	code0
Correct and very precise	Correct	1+points correct· slope adequate	Slope and direction of reflection adequate	Image parallel reflection adequate	Horizontal or vertical slope adequate	Horizontal or vertical image parallel	Horizontal or vertical not slope adequate	Others	Blank

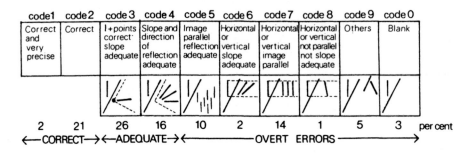

| 2 | 21 | 26 | 16 | 10 | 2 | 14 | 1 | 5 | 3 | per cent |

←—CORRECT—→←—ADEQUATE—→←—————————OVERT ERRORS—————————→

Fig. 10.1   Example of the marking scheme showing percentage of 14 year olds giving each type of response (for item A1.8).

In the analysis of the test the code 1 and code 2 answers were usually combined, and together regarded as 'correct'. The code 3 and code 4 answers are less accurate but can be described as 'adequate' in the sense that they contain none of the 'overt errors' of the remaining codes. These three terms will frequently be used in what follows, with the same meanings as apply here.

The reflection items

The first part of the test was concerned with single reflections, and was made up of six questions containing altogether 27 parts or items.

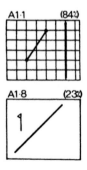

The first question, A1, contained 9 items. Here children were asked to draw freehand the images of single points, straight lines (some with a flag-head) or, in one case, a triangle. The items spanned a wide range of facility, for example, A1.1 was very easy whereas only a minority (of 14 year olds) succeeded with A1.8, which by contrast was on plain paper with a slanting mirror-line (see diagram).

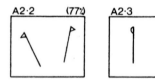

In Question A2 (4 items) children were asked to draw mirror-lines between pairs of figures, as in A2.2, or to state that this was not possible, as in A2.3, where a substantial proportion of children (26 per cent) drew vertical or diagonal lines between the two flags.

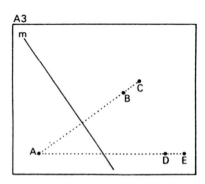

In question A3 (2 items) children had to choose which of 4 given points was the image of the point A (84 per cent) and then explain their choice. By asking for an explanation it was hoped that children would be encouraged to think *analytically* about the properties of reflection, even if they had originally made an intuitive choice. In the event, only 21 per cent gave an explanation that referred to the distance *and* the direction of the point in relation to the mirror-line. Most children focussed on only one of these properties, with 33 per cent choosing distance and 20 per cent direction. Interestingly, very few children used terms like 'perpendicular', '90°' or 'at right angles', instead the direction was described as 'straight across' or 'directly opposite', etc. Another 13 per cent gave tautological answers such as 'point B is where the line m would reflect point A' or 'if you fold it, A goes to B' (note that reflection had been introduced through the idea of folding on the trial items).

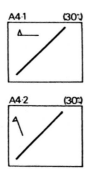

Question A4 (2 items) was similar to A1, except that children were asked to draw their answers carefully and to use a ruler. After the varied experiences of Question A1 and the opportunity to focus explicitly on the rules governing reflection in A3, the question should be a good indicator of children's understanding.

In Questions A5 (7 items) and A6 (3 items) children were asked for the coordinates of a number of single points, after reflection in a vertical mirror-line. The purpose of the question was to see whether children could cope with reflections when the object, image or mirror-line were off the page so that the items had to be answered analytically. The reflections themselves were extremely simple and the easiest item had a facility of about 90 per cent, but this fell to about 40 per cent when part of the transformation was off the page.

Children's errors on the reflection items

Nearly all the children tested showed some understanding of reflection. However, their performance on a given item was heavily dependent on the presence or absence of certain features, which are described below.

The slope of the mirror-line

Children found it much easier to cope with a vertical (or horizontal) mirror-line than one that was slanting, which can be seen by comparing the facilities (of the 14 year olds) shown in the diagram overleaf.

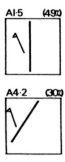

A common error when the mirror-line was slanting was to ignore its slope and simply reflect across or down the page (see Fig. 10.2). This occurred particularly when the object was complex (a flag rather than a single point) and when the slope of the object was itself horizontal or vertical, as in A4.1. Furthermore, for this item almost all those who reflected horizontally also drew the flag parallel to the object, whereas virtually none did so for a slanting flag, as in A4.2. (Given these differences between A4.1 and A4.2, it is interesting to note that the proportion of children answering the items *correctly* was exactly the same, which suggests that once children have reached a certain level of understanding they are more or less immune to these distracting features.)

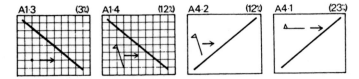

Fig. 10.2   Percentage of children reflecting horizontally (or vertically).

The presence or absence of a grid

As might be expected, a grid can be a powerful guide to the choice of distance and direction in which to reflect an object. This can be seen from the facilities in the diagram below.

However, for the two items shown, the percentage making overt errors (e.g. reflecting horizontally) was virtually the same (6 per cent and 7 per cent respectively), which suggests that the presence of a grid does not necessarily help children overcome such errors.

Complexity of the object

An increase in the complexity of the object can have a marked effect on facility and also on the quality of children's answers. In the case of A1.3 and A1.4, for example, the drop in facility was substantial (see diagram).

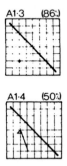

Moreover, many of the children who could cope with the single point and who might therefore have been expected to reflect at least one end-point of the flag correctly, were unable to do so. Only 12 per cent of the children gave adequate answers in A1.4, which hardly makes up for the 36 per cent drop in facility. Also 15 per cent gave answers that were completely disorientated, which is substantially higher than the percentage making *any* overt errors in A1.3.

The slope of the object

Comment has already been made on the tendency to reflect horizontally or vertically and on the fact that this is particularly strong when the slope of the object is itself horizontal or vertical. The same is true for the tendency to draw the image parallel to the object, as can be seen by comparing the frequencies in Fig. 10.3.

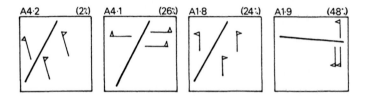

Fig. 10.3   Percentage of children drawing image parallel to object.

Children's understanding of reflection

The previous two sections gave a brief survey of the reflection items and their facilities, and described children's overt errors. An attempt will now be made to interpret and order these empirical findings by examining some of the concepts involved. In particular it is intended to look at the problem of controlling the angle between the object and the mirror-line, which provides a useful focus for discussing these concepts.

Conservation of angle

As can be seen from the frequencies in Fig. 10.4 below, nearly all the children tested gave at least adequate answers to items A1.1, A1.3 and A1.7. The common feature of these three items is that only one slope has to be controlled — of the object in A1.1, and of the mirror-line in A1.3 and A1.7. (Given the primitive notion that a reflection takes the object to the other side of the mirror-line, there is no reason in A1.1 why the direction in which the object is moved should be anything but horizontal so that the specific slope fo the mirror-line can be ignored.)

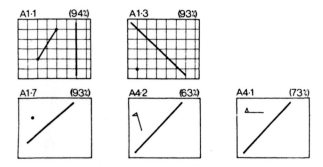

Fig. 10.4   Percentage of children giving answers that were adequate or correct.

On the other hand, substantially fewer children were able to cope at least adequately with items like A4.1 and A4.2, where it is necessary to take note of two slopes (object and mirror-line), and where, furthermore, these slopes have to be *coordinated* so as to conserve the angle between them. (In A4.1, for example, it was common for children to ignore the mirror-line entirely or to consider this slope quite separately from the slope of the object, as is shown in Fig. 10.5).

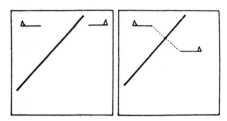

Fig. 10.5   Non-coordination of slopes.

The need to coordinate slopes effectively partitions the items into two groups with respect to both the *qualitative* understanding required (to produce adequate answers) and the *quantitative* understanding required (to produce correct answers). To cope qualitatively with the simpler group (single slope) the child need know little more than that a reflection takes the object to the other side of the mirror-line or that the image slopes 'the other way' (referring back to A4.1, the latter means that a horizontal flag will stay horizontal if it is judged independently of the mirror-line). Both these acquisitions are only a small step removed from the physical act of folding.

Quantitatively, this simpler group requires control over just two aspects which can be carried out in a step-by-step manner. For a single point and a slanting mirror-line (A1.3 and A1.7) these aspects are the direction in which the point is moved followed by the distance moved, whilst for a line (or flag) and a vertical mirror (A1.1) the image can be constructed by first locating one of its end-points (which is just a problem of controlling distance since the mirror-line is vertical) and then drawing the rest of the line (which involves controlling its slope, but which is made easy by the presence of the grid). On the other hand, to cope quantitatively with items involving a flag and a slanting mirror-line (A4.1 and A4.2) a step-by-step approach is unlikely to be sufficiently precise. Rather it would seem necessary to locate each of the end-points separately (they can subsequently be joined). This approach is far more *analytical* and would seem to require a higher level of understanding since it shows a wiillingness to abstract individual elements from the object and since discrete steps are involved which need to be 'stored' while the next is carried out.

Levels of understanding of reflection

Since children's performances can be heavily dependent on the nature of the reflection items, it is not sensible simply to ask whether or not children understand reflection, understanding (both qualitative and quantitative) would seem to be a matter of degree. In support of this it has been shown that certain features can have very different effects on different children. For example, whilst some of the children who coped with a single point (A1.3) could also cope with a second point (A1.4) others could cope with neither, and whereas the presence of a horizontal instead of a slanting flag (A4.1 and A4.2) had a pronounced effect on the type of errors made it had no effect on the number of children answering the items correctly.

In the light of all this it was decided to classify children's understanding into a number of levels. These levels were constructed by grouping items (from the test as a whole) in accordance with some of the methods outlined in Chapter 1.

Generally, children were assessed as being at a given level if they answered the appropriate items correctly — which meant that they could usually cope *qualitatively* with items at a higher level. This is illustrated in Table 10.1, which shows (if a proportion of two-thirds is taken as a criterion) that children at *all* levels (including those children in the total sample who were assigned to Level 0) could cope qualitatively with items that did not need the coordination of slopes. Moreover, it can be seen that (qualitatively) this coordination is acquired by children at Level 2 even though the relevant items are mostly at Level 4.

Table 10.1 Percentage of children at each level (using total sample, n $\simeq$ 1000) who gave at least ADEQUATE answers to items from different levels

Item Level		L0	L1	L2	L3	L4	L5	Need to coordinate slopes
				Level of children				
A1.1	L1	75	97	99	99	100	98	No
A1.3	L1	80	93	98	99	99	99	
A1.7	L2	73	93	99	98	98	100	
A1.4	L2 +	23	38	81	84	100	98	Yes
A4.1	L4	26	44	72	78	100	98	
A4.2	L4	44	63	81	82	99	96	

The levels will now be discussed in detail.

Level 0

Most of the children in the total sample who were classified as being at Level 0 appeared to have internalised reflection sufficiently to be able to cope at least 'adequately' with items that involved single points or a vertical mirror-line (A1.1, A1.3, A1.5, A1.7) where there is no need to coordinate slopes.

Level 1

Children at Level 1 were still generally unable to coordinate slopes. However, they could cope accurately with items A1.1 and A1.3 where it is necessary to exercise deliberate control over two aspects in sequence (position and slope in A1.1 and direction and distance in A1.3). This is a genuine advance over the largely qualitative approach of Level 0 even though the amount of control needed is minimal as the items are on a grid.

At the same time, children at Level 1 were unable to exercise this control in a critical manner. For example, in A1.7, which is without a grid, the most common wrong answer given by the children at Level 1 (37 per cent) was of the type shown in the diagram below, where it can be seen that the tendency to reflect directly across the page has only been partially overcone.

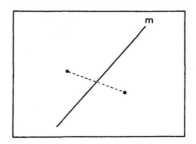

This lack of critical awareness can also be seen in Level 1 children's responses to Question A2, in which they were asked to draw a mirror-line between pairs of figures or, if this is not possible, to say so. A2.2 is a Level 1 item which can be solved in two successive steps by locating a point mid-way between corresponding points of the flags and then drawing a line that is kept mid-way between them.

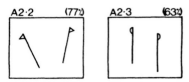

On the other hand, A2.3 is at Level 2. Here Level 1 children, instead of recognising that there is no mirror line, tended to draw lines as in the diagram below. These lines bisect the space between the flags but clearly do not map one flag onto the other.

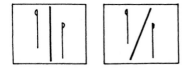

Fig. 10.6   Common Level 1 responses to A2.3.

There were 10 Level 1 items altogether of which A1.1, A1.3 and A2.2 have just been discussed. Of the others, A1.2 was almost identical to A1.1 except that the whole diagram, including the grid, has been turned slightly. Another 4 items were from Question A5 and involved simple reflections, none of which went off the page, but where the answer had to be expressed in coordinate form. The two remaining items involved rotations.

Level 2

Children at this level made two advances over those at Level 1. The first was the ability to coordinate slopes (though without being able to perform such reflections accurately). Second, for items that could be solved by controlling two aspects in sequence, children at Level 2 seemed able to exert this control in a critical manner so that, in effect, they could recognise the inadequacy of the Level 1 answer to A1.7 discussed above and the contradiction inherent in drawing a mirror-line for A2.3.

This critical awareness was also apparent in their answers to A1.6 (triangle and vertical mirror-line) which was classified as a Level 2 item, but for which children only had to locate one point correctly and draw an adequate slope for each side. Typically, children built up the image step-by-step by locating one of the vertices and then adding two of the sides as in the diagram overleaf, with the result that the third side (shown dotted) was often drawn sloping the wrong way.

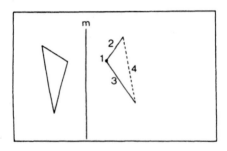

Fig. 10.7   Common method of building up the image in A1.6.

There were 8 Level 2 items altogether, which in addition to A1.6, A1.7 and A2.3, included two further reflection items which might more accurately be placed at the transition between Levels 2 and 3. One of these (A1.4) involved a slanting flag and slanting mirror-line but was on a grid so that there was a greater possibility of answering the item correctly using a step-by-step approach. In the other (A5.4) children were asked for the coordinates of the image of the point (2.1) after reflection in the line x = 12; both features were shown on a diagram so that the distance between them could be measured or counted, even though the image went off the page.

Level 3

There was no substantial difference between children at Levels 2 and 3 in their ability to cope with reflections involving a flag and a slanting mirror-line. However, children at Level 3 could cope accurately with reflections where the object or the mirror-line were off the page so that the distance between them had to be derived analytically rather than directly from a diagram (A5.7, A6.2, A6.3). The remaining two Level 3 items involved rotations.

Level 4

With the possible exception of A1.9 where the mirror-line was just off the horizontal and children tended to exaggerate the slope of the image (which was counted as 'correct' for this item if one of the end-points was located correctly), children at Level 4 could cope accurately with items that involved flags and slanting mirror-lines and that would seem to require a fully analytic approach.. There were 7 Level 4 items of which A1.8, A1.9, A4.1 and A4.2 involved reflections.

Level 5

There were no single-reflection items at Level 5 and there was no clear difference between children at Levels 4 and 5 in their ability to cope with single reflections.

Summary (reflections)

Put briefly, the Level 1 and Level 2 reflection items involved single points or vertical mirror-lines and could be answered correctly by controlling just two aspects in sequence. The main distinction between the two Levels is that the Level 1 items were on a grid. The Level 3 items involved reflections that were partly off the page whilst the items at Level 4 required the coordination of slopes and were unlikely to be answered correctly unless a fully analytic approach was used.

In Piagetian terms, it can be hypothesised that the items at Levels 1 and 2 require concrete operational thought as they are very much tied to the activity of folding. On the other hand, the need to cope with reflections that go off the page and generally to use an abstract and analytic approach suggests that the Level 3 and 4 items require at least the beginning of formal operations.

The diagram below gives examples of single-reflection items at Levels 1—4 together with their facilities for the 14 year olds (the facilities of all the items used to construct the levels are listed towards the end of this chapter for each age-group tested).

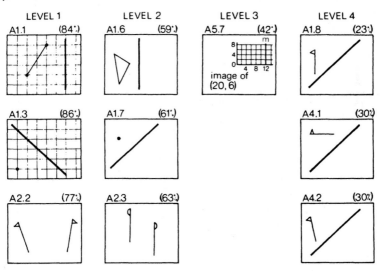

Fig. 10.8   Examples of reflection items at each level.

The rotation items

Section B of the test was concerned with single rotations and consisted of 4 questions made up of 21 items (with an additional 4 trial items).

In question B1, which consisted of 11 items, children were asked to sketch the images of various flags or triangles after rotations of a quarter turn (anticlockwise). Some of the items were on plain paper and some on a grid. Their facilities varied considerably, depending on the position of the centre of rotation and the slope of the object (and, to a lesser extent, the object's complexity and the presence or absence of a grid). Thus, of the items illustrated below, 85 per cent of 14-year-old children answered B1.1 correctly, in which the centre is on the flag and the flag itself is vertical, whereas only 17 per cent answered B1.11, which involves a slanting flag not passing through the centre.

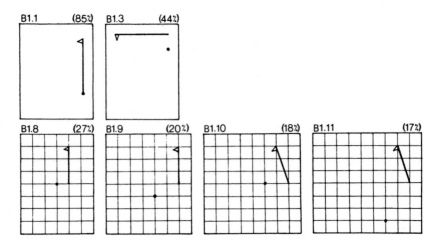

With the exception of B1.10, children found it easier to determine the correct slope of the image than to locate its correct position. For example, in item B1.8 35 per cent drew an image with the correct slope (but in the wrong position) whilst only 1 per cent located the base-point of the flag correctly (but with the wrong slope). For B1.10 the corresponding percentages were 9 per cent and 11 per cent, but here the flag is slanting rather than vertical whereas the line from the centre of rotation to the base-point of the flag is horizontal. (Note: these percentages do not include children who may have drawn one aspect of the image correctly but who made extreme errors such as drawing the image through the centre when originally the object did not pass through the centre. In general this extreme category will be ignored when discussing specific features of children's responses).

The items on a grid tended to be more reliable than those on plain paper in the sense that their correlations with other items were usually higher. However, in contrast to the reflection items, the presence of a grid did not necessarily make the items easier, as can be seen by comparing the facilities of B1.3 and B1.8, shown above. It seems that the grid encouraged children to place the image 'symmetrically' with respect to the centre by drawing the base-point P at P^1 (or P^{11}) as shown in the diagrams below. Such answers were given by only 16 per cent of children for B1.3 but by 32 per cent and 25 per cent for B1.8 and B1.9 respectively.

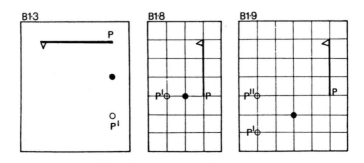

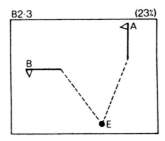

Explain why E is NOT the centre of rotation

Question B2 consisted of 3 items, in each of which children were given a drawing of a flag and its image after a rotation of a quarter turn, together with a point joined to the base-points of the two flags by dotted lines, as in this diagram. The children were told whether or not the point was the centre of rotation and were asked to explain why. Forty-nine per cent could do so when the dotted lines were of different lengths (item B2.1, not shown) but only 23 per cent could do so in B2.3 where the lines were of the same length but not at right angles.

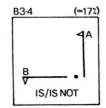

B3 also consisted of several drawings of a pair of flags and a point, but this time children were simply asked to state whether or not the point was a centre of rotation. (The question consisted of 4 parts but was marked as a single item, so that the exact facilities of the separate parts is not known.)

B4 consisted of 6 items in which the children were asked to draw centres of rotation. The easiest of these was B4.1 and the most difficult B4.4 in which 21 per cent drew the centre at the point X shown below. B4.5 was similar to the last part of B3 and here 42 per cent drew the centre at the point Y with another 9 per cent choosing a point on the dotted line (but not such that the lines from their centre to the base-points of flags were at right angles).

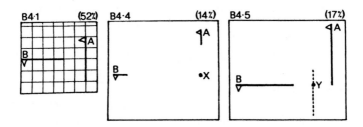

Levels of understanding of rotation

Level 1

The two rotation items at Level 1 involved a vertical flag (B1.1, 85 per cent) and an L-shaped flag, one part of which was horizontal the other vertical (B1.2, 74 per cent). In both cases the centre of rotation was at the base-point of the flag. To answer these items correctly it was necessary to rotate the flags through a quarter turn, but the actual size of the images was ignored in the marking scheme. Thus children at Level 1 could control the change in slope of the object, even if this involved two steps as in B1.2, given that this change was from horizontal to vertical, or vice versa, and given that the centre was on the object. However, these children could not generally cope with a more complex object such as a triangle (B1.6, 67 per cent) which they tended to rotate through an arbitrary angle (not a quarter turn) or to turn over, nor could they cope with rotations where the centre was not on the object.

Level 2

The Level 2 rotation items were B1.6 (67 per cent) which involved a quarter turn of a triangle with the centre at one of the vertices, B2.1 (49 per cent) which is shown in the diagram below, and B3 (61 per cent), for which it was necessary to answer 3 out of 4 parts correctly.

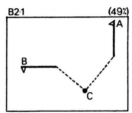

Explain why C is NOT the centre of rotation

To solve B2.1 children had only to recognise that the two dotted lines (which were given) were not the same length. This was also sufficient to answer the 3 easier parts of question B3 (but not the last part, in which children were given a centre midway between the end-points of a flag and its image). Thus one of the main achievements of children at Level 2 was that they could recognise that the centre of rotation had to be equidistant from a pair of corresponding points on the object and image. However, they did not generally realise that this had to be true for all (or more than one) such pairs of points nor were they aware that for a quarter turn the lines joining the centre to such points had to be at right angles (or, put another way, that a line from the centre to any point on the object itself went through a quarter turn).

This achievement, and its limitation, is also illustrated by the fact that 69 per cent or about two-thirds of the children at Level 2 in the total sample were able to locate a centre equidistant from the end-points of the flags in B4.5 (shown in the diagram) though this is a Level 5 item. However, the point chosen was usually (51 per cent of all Level 2 children) drawn midway on the line joining the end-points.

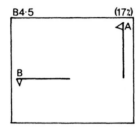

A second, and in many ways remarkable achievement of children at Level 2 was the ability to draw images with a correct slope (though not usually in the correct position), given that the original flag was horizontal or vertical. This can be seen from the Table below.

Table 10.2   Percentage of children at each level drawing image with correct slope (but not necessarily correct location). (Proportions less than two-thirds underlined.)

	Item	Level of Children					
	Level	L0	L1	L2	L3	L4	L5
B1.3	L3	17	50	71	89	98	100
B1.8	L4	17	50	74	82	96	99
B1.9	L5	16	40	68	77	89	99
B1.10	L5	4	5	21	32	59	93
B1.11	L5	2	3	15	35	52	95

Level 3

There were two rotation items at Level 3. In both, children were asked to rotate a flag through a quarter turn but with the centres not on the flags. One of these involved a horizontal flag where the centre was directly underneath the end-point (B1.3, 44 per cent) and the other involved a slanting flag which was in line with the centre (B1.4, 41 per cent). Thus the major breakthrough of children at Level 3 was the ability (in simple cases) to perform rotations which required, at least mentally, that a line be constructed from the centre to the object. However, these children were not able to perform such rotations in the presence of a grid, which seems to act as a distractor when it comes to locating the position of the image.

Level 4

There were three rotation items at Level 4. One of these (B1.8, 27 per cent) is similar to the Level 3 item B1.3, but with a grid. Here the flag is vertical and the line from the centre to the base-point of the flag is horizontal. In another of the items (B2.3, 23 per cent) children have to explain why a point which is shown to be equidistant from the base-point of a flag and its image is not the centre of rotation. Thus children at Level 4 were beginning to recognise that the angle between the lines from the centre to corresponding points on the object and image is equal to the angle of rotation (as well as the lines being of equal length, which was recognised at Level 2). However, these children were still not generally able to use this understanding to perform rotations unless the lines and the slope of the flag were horizontal or vertical.

Level 5

There were seven Level 5 items involving single rotations. In three of these (B1.9, 20 per cent, B1.10, 18 per cent; B1.11, 17 per cent) either or both the flag and the line from the centre to the base of the flag were slanting. A fourth item (B1.7, 17 per cent) involved a rotation of a triangle with the centre inside it, where it was necessary to control the relative distances from each of the sides as well as drawing the slopes of the sides correctly. In the remaining items (B4.4, 14 per cent; B4.5, 17 per cent; B4.6, 20 per cent) children had to find the centres of (quarter turn) rotations. Thus children at Level 5 could rotate any flag through a quarter turn regardless of its slope or its position relative to the centre. In particular, these children could *use*, as well as recognise, the fact that any pair of lines from the centre to corresponding points on the object and image form a right angle.

Summary (rotation)

At Level 1 children could draw an image with the correct slope if the object was horizontal or vertical and the centre on the object. At Level 2 the centre could be anywhere, but it was not generally until Level 5 that children could control the slope if it was slanting.

As far as *position* was concerned, children at Level 2 began to recognise that the lines from the centre to corresponding base-points were the same length, but not that they were at right angles. At Level 3 children could actively construct such lines and use them successfully if they were horizontal or vertical (or if the centre was on the extension of the object) and not on a grid. The distraction of a grid was overcome at Level 4 and these children began to recognise that such lines were at right angles even if they were slanting. However, this did not become fully generalised and operational until Level 5.

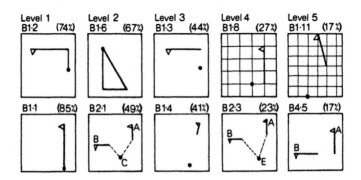

Fig. 10.9   Examples of rotation items at each level.

Reflections and rotations combined

Section C of the test was concerned with combinations of reflections and rotations and was made up of six questions, of which four were used in the final analysis of the test. These questions were marked as single items and were all classified at Level 5.

In one of these (C5, 13 per cent) children had to apply the inverse of a given rotation in order to find an unknown transformation which, followed by the rotation, mapped a given flag onto another given flag.

In the other three items (C1, C2 and C3) children were asked to draw the image of a given flag after applying two transformations in sequence, and then draw a mirror-line or a centre of rotation to represent the equivalent single transformation. (Answers were only classed as correct if this last step was correct.)

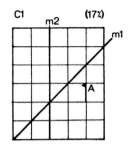

In the case of C1 the two transformations were reflections in the lines m 1 and m 2 as shown in the diagram. Children found the question extremely difficult (it was answered correctly by only 17 per cent of 14 year olds) but this seemed to be due as much to the difficulties involved in applying the transformations as in identifying the resulting transformation. Thus only 32 per cent of the 14-year-old children and generally only the children at Levels 4 and 5 in the total sample (taking a criterion of two-thirds) were able to draw the correct final image of the flag.

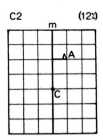

A similar pattern of results occurred for C2. Here the given transformations were a reflection in the mirror-line m (shown in the diagram above) followed by a rotation of a quarter turn about the point C. Only 12 per cent of the 14 year olds were able to identify the resulting transformation, and only 21 per cent of 14 year olds and generally only the children at Level 5 could draw the correct final image.

Question C3 involved the same transformations but with the rotation applied first, and here the corresponding percentages were 11 per cent and 18 per cent (Level 5).

Performances for different year groups

Listed below are the facilities of all the items (reflections and rotations) that were used to construct the levels (Table 10.3), and also the distribution of children at each level for the 13, 14 and 15 year olds (Table 10.4).

Table 10.3   Facilities of all the items (percentage)

Level	Items	13 yrs	14 yrs	15 yrs	Level	Items	13 yrs	14 yrs	15 yrs
Level 1	A1.1	82	84	83	Level 4	A1.8	22	23	33
	A1.2	77	79	82		A1.9	31	28	41
	A1.3	79	86	87		A4.1	25	30	40
	A2.2	73	77	83		A4.2	25	30	39
	A5.1	88	88	91		B1.8	23	27	32
	A5.2	88	81	88		B2.3	24	23	29
	A5.3	77	79	84		B4.2	26	31	35
	A5.5	76	76	81					
	B1.1	83	85	91	Level 5	B1.7	16	17	25
	B1.2	67	74	77		B1.9	14	20	24
Level 2	A1.4	48	50	66		B1.10	14	18	25
	A1.6	51	59	67		B1.11	11	17	21
	A1.7	55	61	65		B4.4	9	14	16
	A2.3	61	63	70		B4.5	13	17	20
	A5.4	46	51	60		B4.6	14	20	16
	B1.6	56	67	71		C1	15	17	22
	B2.1	52	49	54		C2	10	12	18
	B3	52	61	63		C3	8	11	15
Level 3	A5.7	37	42	50		C5	12	13	20
	A6.2	31	36	44					
	A6.3	37	45	50					
	B1.3	37	44	46					
	B1.4	36	41	45					

Implications for teaching

For combinations of transformations the results indicate that if children are to cope with these at all, the constituent transformations will have to be very much simpler than the ones used in the test. This is supported by the findings of Collis (1975) and Sheppard (1974) who used cardboard shapes to train children to perform and recognise actions such as 'turning over' and 'turning round' which mapped the shapes onto themselves. It seems likely that even children below Level 1 would be able to cope singly with such transformations, which are essentially qualitative, and Collis and Sheppard found that 'average 10 year olds' could readily identify their

Table 10.4   Distribution of children at each level (percentage)

	13 yrs	14 yrs	15 yrs		13 yrs	14 yrs	15 yrs	
Level 0	17	16	12	Level 4 5/7 items	8	8	12	
Level 1 6/10 items	30	25	20	Level 5 7/11 items	4	8	11	
Level 2 5/8 items	17	16	19	Total number of children	293	449	284	1026
Level 3 3/5 items	23	28	26					

combinations, but only when the transformations were of the closed 'turning round' type.

For the single transformations, an encouraging aspect of the results reported in this chapter is that nearly *all* the children tested had some understanding of reflection and rotation, which means a basis exists for studying the transformations in secondary schools. At the same time, control over these transformations posed substantial conceptual problems to most children (which were far more severe than had originally been predicted), so that a study, even of single transformations, need be far from trivial.

If the transformations are to be studied in their own right, it would seem pointless to do this in a didactic, expository manner. The fact that the transformations can be defined in terms of actions (folding and turning) and their results represented in a very direct manner by drawings means that the topic is ideally suited to a practical and investigative approach. The actions and the representations are both highly intuitable so that it should be possible to develop such an approach in ways that are meaningful to most children. The transformations can be internalised in gradual steps, by focusing first on the actions themselves, then on their representation, and then on the representation of imagined actions. In addition, the resulting drawings can be checked at each step by a return to the actions or by reference to drawings of simpler problems (though it should be said that children, in common with the rest of us, may have difficulties in recognising the need for such checks, and when carrying them out may see only what they want or expect to see).

The approach being advocated is one that directs children towards discoveries from which the rules and properties of the transformations can be surmised and against which they can be tested. Such an approach might appear closer to the activities of science than mathematics (without being so cumbersome or messy), but this is all to the good, given the belief that such activities are vital to the development of critical thinking.

11   Vectors and matrices

VECTORS

Virtually all 'modern' courses and syllabi give an important place to matrices and vectors although there is no general agreement about when to introduce the topics.

The level to which vectors are taken is not constant even within courses designed for O-level or courses designed for C.S.E.

Topics included

The following areas were selected from the common core which seemed to exist in all the texts:

Vectors as ordered n-tuples i.e. numerical column vectors either in a data storage 'story' context or without any 'meaning' attached. Simple addition, subtraction and scalar multiplication.

Vectors as ordered n-tuples representing movements and positions i.e. single and double translations, moves in two and three dimensions.

Free vectors i.e. using the notation $\underline{a}$ $\frac{1}{2}\underline{a}$, structure and elementary geometrical applications.

Route vectors i.e. using the notation $\overrightarrow{AB}$, $\overrightarrow{BC}$ for movements.

The results

(14 and 15 year olds were tested)

Vectors as ordered n-tuples

The three questions (1, 2 and 3) in the section dealt with numerical column vectors either storing data or with no 'meaning' attached. Movements and positions were not involved.

Question 1 was an introductory question using column vectors to store information on the parts needed to make types of radio. Addition, subtraction and scalar multiplication were involved as well as the interpretation of such operations. At the interview stage it became clear that firstly the interpretation of such operations proved to be more demanding than the numerical manipulation required and secondly there was a tendency among less able subjects to work out each answer from first principles. The question was so structured that previously found answers had to be added together to give the final answer, but less able subjects rarely did so

158

and worked out the required information again. Facilities ranged from 91 to 64 per cent for 14 year olds and from 93 to 71 per cent for 15 year olds.

Question 2.   In this question no data storage was involved and straightforward addition, subtraction and scalar multiplication of column vectors proved to be easy; facilities were over 85 per cent for both year groups. Adding vectors involving negative integers produced the expected sharp drop in facility to 50 and 70 per cent. The equation $4\begin{pmatrix} 4 \\ b \end{pmatrix} = \begin{pmatrix} a \\ 8 \end{pmatrix}$ was the last part of this question. Two interesting incorrect strategies appeared in the results and on interview.

Table 11.1   Answers for $4\begin{pmatrix} 4 \\ b \end{pmatrix} = \begin{pmatrix} a \\ 8 \end{pmatrix}$ (percentage)

	14 yrs	15 yrs
Correct 'a' value	71.3	79.1
Correct 'b' value	76.1	80.4
Ignoring the scalar multiples and solving $\begin{pmatrix} 4 \\ b \end{pmatrix} = \begin{pmatrix} a \\ 8 \end{pmatrix}$	5.0	3.0
Changing the equations $\begin{pmatrix} 4 \\ b \end{pmatrix} = 4\begin{pmatrix} a \\ 8 \end{pmatrix}$	4.0	2.0

Question 3.   Simple examples of splitting vectors into components (base vectors) were involved here. The word 'base' was not used and no geometrical interpretation was given or required. The three introductory items had fairly high facilities (63.8 to 55.4 per cent for 14 yr olds and 72.9 to 63.3 per cent for 15 yr olds) especially in view of the lack of emphasis given in most textbooks and school courses. The last item asked for the missing numbers to be filled in:

$$\begin{pmatrix} 3 \\ 7 \end{pmatrix} = \quad \ldots \begin{pmatrix} 1 \\ 1 \end{pmatrix} + \quad \ldots \begin{pmatrix} 0 \\ 2 \end{pmatrix}$$

	Success	One value correct
14 yr	44.6	23 per cent
15 yr	55.2	16

On interview many subjects managed to obtain the first missing number, 3, but failed to work out the more difficult 2 indicating that the difficulty lay in equation solving rather than in understanding what was required.

Vectors as ordered n-tuples representing movements and positions

This section forms a logical progression from the previous one and uses vectors to represent movement and positions, 2- vectors for 2 dimensions and 3- vectors for three dimensions (Questions 4, 5, 12).

Question 4.   The map of the island formed the basis of the question. Most of the items involved finding moves and adding chains of moves using vector equivalence to check. The main type of common error observed on interview was the tendency to reverse the axes, i.e. $\begin{pmatrix} 2 \\ 1 \end{pmatrix}$ for $\begin{pmatrix} 1 \\ 2 \end{pmatrix}$ (5 per cent 14 yr olds and 3 per cent 15 yr olds in the survey).

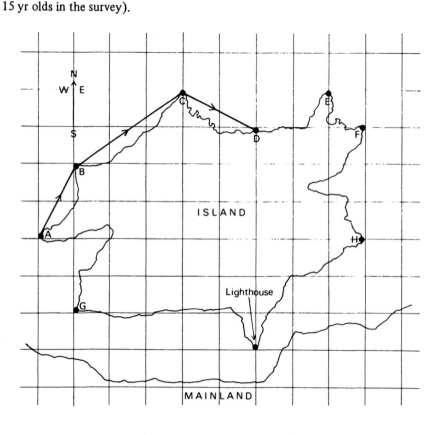

Seventy-eight per cent of each year group could give an adequate reason why C to D was written $\begin{pmatrix} 2 \\ -1 \end{pmatrix}$. Clear evidence was obtained that pupils of this age are willing to accept the use of negative integers providing a concrete reason for using them exists. Typical answers were of the form 'Because you go down not up' or 'Because you are going south not north'. When asked to write down the series of moves D to E, E to F and F to H, the facility of each item was lower than that of the previous item. The need to use negative integers and zeros accounts for this. The facilities are given in the Table below together with the facility for Item 23 which asked children to add the three previous answers and Item 24 which asked for its interpretation.

Table 11.2   Results of items 20—24 (percentage)

	Item 20	Item 21	Item 22	Item 23	Item 24
	D to E	E to F	F to H	Add	Meaning
14 yrs	75.8	63.0	53.9	40.7	22.4
15 yrs	77.5	68.3	59.1	49.2	24.4

Many subjects who failed on $\begin{pmatrix} 2 \\ 3 \end{pmatrix} + \begin{pmatrix} -1 \\ -4 \end{pmatrix}$ or $\begin{pmatrix} -2 \\ -4 \end{pmatrix} + \begin{pmatrix} -1 \\ 5 \end{pmatrix}$ or both, succeeded

on items 21, 22 and 23. The fact that a concrete model exists is clearly important for the acceptance of negative integers.

Question 5.   The question concerned translations by using vectors and was accompanied by the diagram shown below, in which $G \rightarrow G^1$ was given as an illustration.

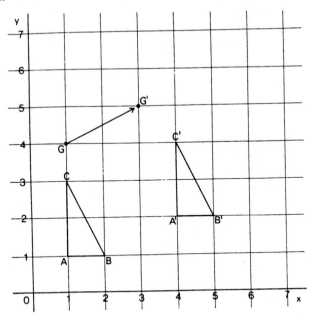

The facilities ranged from 57.7 per cent to 31.9 per cent for 14 yr olds and 61.8 per cent to 32.2 per cent for 15 yr olds. About half of each year group were able to complete a simple translation but the most interesting parts were items 30 and 31.

Item 30 asked 'which vector would translate ABC straight to A″B″C″ (the move from A′B′C′ to A″B″C″ was the previous item) and item 31 asked how this was worked out. The A″ notation gave no trouble and was accepted without question by all the subjects on interview.

Table 11.3 Method of working out a double translation (Item 31)

Count squares	Add vectors	
43.2	14.5	14 yrs
44.1	17.6	15 yrs

The Table above shows the results of item 31, broken down into the two possible types of answer, which were separately coded. Adding the vectors is the more abstract method and children who chose to do so were generally superior on the test as a whole to those who counted squares; more subjects aged 15 chose the more abstract method. Increased willingness to use abstract methods appears to be an important factor in success on more complex vector work, indeed on interview many of the subjects who chose to count squares could not see another method even when prompted. The last item asked for the new position of a transformed line segment but gave no diagram. About one-third of each year group managed to find the new position by adding the translation vector applied to the original position vectors given. As expected the lack of a concrete model produced a drop in facility compared to the previous item. The question was specifically designed to make it extremely difficult to draw a diagram and the assumption can be made that one-third of each year group were able to handle abstract translations.

Question 12. This three dimensional question using 3-vectors proved less demanding than expected. The items concerned the movements of a fly, asking where it would be after certain moves and working out the moves from one position to another. Facilities ranged from 51 to 32 per cent and from 57 to 39 per cent, the 15 yr olds scoring higher.

The final item asked children to reverse the previous move (find its inverse) and proved to be interesting on interview. Some subjects simply inverted the previous move by multiplying the 3-vector by − 1 while others found it necessary to work the move out from first principles. Considering the fact that no model or diagram can be easily constructed or drawn the facilities were rather higher than might have been expected.

Free vectors

This section used the free vector notation a for vectors and contained items on elementary structure and geometric applications (Questions 6, 7, 8, 9, 10).

Question 6. The question introduced the notation involved, showing a, − a and 2a and asked for all the scalar multiples of a to be marked in the diagram shown opposite.

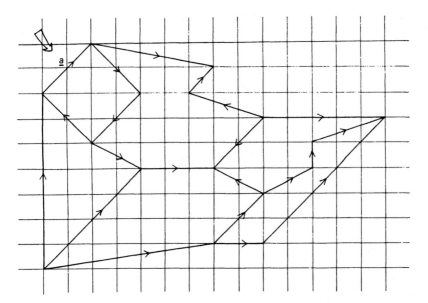

Table 11.4   Marking families of vectors (Item 28)

Parts Correct:	6	≥ 5	≥ 4	≥ 3
14 yrs	15.4	38.8	61.3	72.6
15 yrs	25.8	47.2	66.7	75.1

The most common error made in attempting to mark the six correct vectors concerned the magnitude whereas direction (as indicated by + or —) presented few problems. $2\frac{1}{2}a$ and $\frac{1}{2}a$ gave the most trouble, 42.6 per cent of those aged 14 and 27.6 per cent of those aged 15 made errors in assessing the magnitude of the correct vectors. Of the rest of the vectors in the diagram those 'nearly parallel' to a were marked by 35.0 per cent of 14 yr olds and 20.9 per cent of 15 yr olds. Orthogonal vectors of the same magnitude as a were less of a problem, 17.9 per cent and 12.4 per cent marked them.

A second question, in which children were asked to draw and compare 2c and — c to a given c, was successfully completed by 70 per cent of each year group.

The magnitude of both vectors proved to be easier to describe than the direction and it was noticeable both on interview and in the written answers that the words parallel and opposite did not appear as frequently as expected. While the former is easy to circumlocute, the latter is not and tortuous English often resulted. 'Opposite' can be regarded as an everyday word and the reluctance or inability to use it seems surprising. On the whole the question showed that the basic idea of a 'family' of scalar multiples of a vector is acceptable to the majority of children in this age range.

Question 8.   The question introduced the triangle law of addition and by using two parallelograms asked for the conclusion that addition of vectors is commutative. One error which occurred frequently here was that of drawing or finding the inverse of the correct vector answer:

Table 11.5   The commutativity of vector addition

Parallelograms:	draw $\underline{e} + \underline{d}$	draw $\underline{d} + \underline{e}$	Conclusion	Inverse error	
Items 50, 51 and 52	51.1	39.3	23.0	18	14 yrs
	55.2	41.8	27.5	20	15 yrs

There is a considerable facility drop between obtaining two correct diagrams and concluding correctly about commutativity. On interview the problem seemed to be that some subjects did not see the presence or absence of commutativity as a point of interest or assumed it anyway.

Question 9.   This was a further question on the triangle law requiring answers of the form $\underline{a} + \underline{b}$ and $\underline{a} - \underline{b}$ but comparing these forms when presented in a triangle diagram and in a parallelogram diagram. The general results can be summarised thus: answers of the form $\underline{a} + \underline{b}$ are easier than those of the form $\underline{a} - \underline{b}$. Of the $\underline{a} + \underline{b}$ type answers the one in the parallelogram proved easier ($\underline{b} + \underline{a}$ acts as a check for $\underline{a} + \underline{b}$) but of the $\underline{a} - \underline{b}$ type items the one in the triangle diagram proved easier. While a check on a parallelogram diagram would give $-\underline{b} + \underline{a}$ for $\underline{a} - \underline{b}$, which is not as helpful to the child, it does not seem likely that this is a hindrance. The picture given by the results remains incomplete. The error involving giving the inverse of the correct vector answer was again common, frequencies varied from 36.8 per cent to 14.4 per cent for 15 yr olds and from 45.4 per cent to 21.3 per cent for 14 yr olds.

Question 10.   Using the diagram below the items in this question involved the use of vector equivalence and forming chains of vectors. On the whole the difference in performance between the two year groups was greatest on the questions involving one vector as the sum of others.

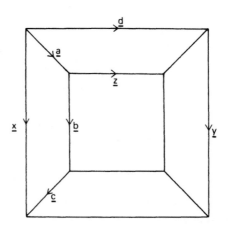

Table 11.6 Facility levels for Question 10

	Marking equivalent vectors	Vector as a sum of other vectors
14 yrs	31—42 per cent	8—34 per cent
15 yrs	46—57 per cent	16—45 per cent

Route vectors

Question 11. The question used the same diagram as Question 10 but the letters were changed to use the $\overrightarrow{AB}$ notation for a vector. The facilities for forming chains of equivalent vectors were high, around 60 per cent. The most interesting part contained three items requiring the use of this vector notation to provide routes but without a diagram, i.e. only letters were provided, see Table 11.7.

Table 11.7 Route vectors with no diagram available (percentage)

		14 yrs	15 yrs
Item 71	$\overrightarrow{SJ} + \overrightarrow{JT} + \overrightarrow{TF} = \ldots$	50.4	57.1
Item 72	$\overrightarrow{SJ} + \ldots = \overrightarrow{SZ}$	29.3	43.5
Item 73	$\overrightarrow{LS} + \ldots + \overrightarrow{TP} = \overrightarrow{LP}$	36.3	47.7

The vectors hierarchy

The items in each group are described in general terms.

Group 1: Elementary work with ordered n-tuples not representing movement or positions
(i) The use of ordered n-tuples to store data in a story context, in particular the addition of n-tuples and their interpretation, (ii) the interpretation of $a\underline{x}$ as a lots of $\underline{x}$ where $\underline{x}$ was an ordered n-tuple.

Group 2: Ordered n-tuples. Further work including elementary movements
(i) The formation and interpretation of scalar multiples of ordered n-tuples storing data in a 'story' context, (ii) the solution of equations involving scalar multiples of ordered n-tuples, (iii) an explanation of the need to use negative integers when writing down vector moves, (iv) writing a move in the form $\begin{pmatrix} a \\ b \end{pmatrix}$ where a and b are positive integers.

Group 3: Elementary geometrical work with vectors
(i) The expression of a given n-tuple in terms of $\begin{pmatrix} 1 \\ 0 \end{pmatrix}$ and $\begin{pmatrix} 0 \\ 1 \end{pmatrix}$ (help given with lay-out), (ii) the writing of a move with form $\begin{pmatrix} a \\ b \end{pmatrix}$ where one element is a negative integer, (iii) a description of elementary scalar multiples of a given vector $\underline{a}$ such as $\frac{1}{2}\underline{a}$, $2\underline{a}$ and $-\underline{a}$ in terms of the magnitude and direction of $\underline{a}$, (iv) the completion of

a simple vector triangle of addition where the vectors are already nose to tail and a similar example has been given, (v) the writing of a simple route vector chain as a single route vector and the expression of one route vector as a chain of others using a given diagram.

Group 4: Elementary abstract movements

(i) The expression of a given n-tuple in terms of $\begin{pmatrix} 1 \\ 0 \end{pmatrix}$ and $\begin{pmatrix} 0 \\ 1 \end{pmatrix}$ and the expression of an n-tuple in terms of other bases when only simple simultaneous equations are involved, (ii) moves of the form $\begin{pmatrix} 0 \\ -b \end{pmatrix}$, (iii) the drawing of the result of a simple translation, (iv) finding the vector for a double translation, (v) working out the result of a translation followed by its inverse where no diagram is available, (vi) the replacement of a string of vectors by one resultant vector when no diagram is provided, (vii) the marking of equivalent vectors when no distractors are present.

Group 5: Further abstract movements

(i) The representation of three successive movements on a diagram as an ordered n-tuple together with the addition of such movements (including negative integers), (ii) the vector for a double translation, (iii) the marking of scalar multiples of a given vector given a selection of vectors, (iv) the marking of equivalent vectors when strong distractors are present, (v) the provision of missing route vectors for a chain, given the resultant but no diagram, (vi) the finding of the inverse of a three dimensional move (no diagram).

Group 6: Explanation, checking and commutativity

(i) The interpretation of the sum of three vector moves on a given diagram, (ii) finding a route and checking it by using vector equivalence, (iii) the explanation of the commutativity of vector addition, (iv) vector equivalence explained in terms of necessary and sufficient conditions.

Group 7: Full use of vector equivalence

(i) The use of vector equivalence in either free vector or route vector notation (difficult items), (ii) the use of vector addition and vector chains involving vectors such as $-\underline{a}$.

Percentage of subjects at each level

Table 11.8   Performance by age groups (percentage)

Level	Overall facility (pass mark)	15 yrs	14 yrs
0		9.3	9.3
1	88–92(3/4)	10.7	15.7
2	75–82(5/7)	16.5	19.1
3	66–72(9/12)	17.9	19.8
4	51–61(7/9)	13.0	16.4
5	34–48(9/12)	18.2	12.2
6	22–28(4/5)	9.9	5.2
7	5–19(3/4)	4.5	2.3

Note: 7.6 per cent error types recoded.

Commentary

The hierarchy can be seen as containing two important threads, abstraction and complexity. With regard to abstraction the children from Level 4 upwards are capable of performing increasingly complex tasks without any concrete representation to help them. The frequency of the comment 'no diagram available' bears testimony to this and the children are increasingly able to withstand the effects of any distracting influences.

An increasing ability to form conclusions and to explain and interpret is observable. The relationship between complex vector work and the ability to cope with abstraction is displayed — the more complex the vector work being undertaken the greater the need to think abstractly, to work without the aid of diagrams and to manipulate vectors without having concrete representations of them.

The older group is superior to the 14 year olds at all levels, the difference being most marked among the higher levels. The relationship between type of response pattern and abstraction is well demonstrated by the choice of method when asked to work out the vector for a double translation as shown in Table 11.9.

Table 11.9   Types of response to Item 31
(percentage)

Level of child	Wrong or no method	Count squares	Add vectors
0	91	9	0.5
1	80	19	1
2	62	35	4
3	44	51	5
4	7	73	20
5	4	62	35
6	1	54	45
7	2	28	70

A sharp change is observable when Level 4 is reached.

Implications for teaching

The mean number (percentage) of correct answers (out of 76) achieved by the two year groups, 38.7 by the 14-year-old group and 43.2 by the 15-year-old group, indicates that much of the vector work described is within the capabilities of the average child of this age. On the whole the subjects did not have a great deal of experience in vector work other than ordered n-tuples, and greater familiarity should make more available to the average child, but the relationship between vectors and abstraction shows clearly in the test and suggests that for the average child concretely based vector work is the most suitable.

Vectors as ordered n-tuples seem to be the easiest introduction to vector work, the importance of ordered data storage can be stressed and such work may be used to give practice in numerical manipulation in more varied contexts. Experience of interpreting and ordering data can also be provided in 'problem' type questions. The facilities of the items concerned indicate that such work is within the capabilities of all but the least able children.

Following this, the use of ordered n-tuples to represent movements and position seems logical and has several advantages. The necessity for an agreed order of presentation when axes are involved can be shown and the willingness of children to accept the need for negative integers in this situation has been demonstrated as has the advantage of providing a concrete situation where the addition of negative integers is needed.

Translation proved rather more demanding than might have been expected and it seems likely that the topic rather than being used to introduce vectors should follow the work described above. Double translations were particularly difficult.

The elementary concept of a free vector is readily understood by third and fourth year pupils (ages 14—15) and seems to be the next logical step. In particular such work can be used to strengthen ideas of parallelism and oppositeness of sense, which appear to need more emphasis. Further work on free vectors, vector chains and triangles of addition proved to be demanding and the marked difference between the performance of the two year groups suggests that this work is better left until the fourth or even the fifth year in school.

The case for the inclusion of route vectors cannot be advanced on the results of the test. Elementary work using the $\overrightarrow{AB}$ notation proved relatively easy but by stressing the end points the notation does not encourage the concept of vector equivalence. More may be achieved with the free vector notation before a high degree of abstraction becomes essential.

In conclusion the test results support the suitability of much vector work for the third and fourth year pupil providing that the level of abstraction required is moderate. More advanced free vector work appears to be more suited to the fourth year of the mathematics course. In the author's view much of value can be achieved without taking vectors to the level required for vector proofs ($\lambda \underline{a} + (1 - \lambda)\underline{b}$ for a point dividing a line in a given ratio for example). The replacement of parrot learned Euclidean proofs by parrot learned vector proofs should not be the result of the inclusion of vectors in the school syllabus.

MATRICES

Introduction

The initial step, a survey of commonly used text books revealed rather more agreement about presentation of matrices than is the case with vectors. The central feature of most approaches is the stress laid upon matrix multiplication. Addition is rarely heavily emphasised and is often mainly considered in the vectors sections. This emphasis is understandable since most matrix applications have multiplication as the chief operation rather than addition. Applications of matrices dealt with in books are many and varied but several general points of interest were identified.

Transformation geometry is an application found in all the books surveyed, while networks seem more favoured in texts designed for C.S.E. courses. Data storage (item × price or position × points given for position in a sporting context), vies with transformations as a vehicle for the introduction of matrix multiplication. Considerable variations in time and order of presentation of different aspects of matrices were observed but in most cases the desire to use matrices as a linking structure between various aspects of mathematics is clearly demonstrated, at least to the teacher.

As a result of the survey the topic areas shown below were selected for possible inclusion in the test.

Topics studied

A   The use of matrices to store data in a 'story' context and the addition, subtraction and scalar multiplication of such matrices.

B   The manipulation of abstract (no 'story') matrices of moderate order including the commutativity of matrix addition.

C   Elementary equations involving the addition and scalar multiplication of matrices containing some unknowns.

D   The logic of data storage including consistency of position and consistent units of measurement.

E   The introduction of matrix multiplication using a data storage approach.

F   The use of matrices to represent networks.

G   The introduction of matrix multiplication using networks.

H   The multiplication of abstract (no 'story') matrices including order and the non-commutativity of matrix multiplication.

I   The use of matrix multiplication to answer 'story' type problems.

J   Writing a 'story' to make use of a given correct matrix multiplication.

K   The relationship between the elements of two matrices being multiplied and the link between this and what the resultant matrix represents.

L   The use of matrices to perform rotations and reflections of a geometrical form.

Having established the impracticability of using one test both for children previously introduced to matrix multiplication and for children not already familiar with the topic, two tests were written: one test (M) for those already familiar with matrix multiplication and another (A) for children for whom multiplication was a new topic.

As many of the same questions as possible were included in both versions of the test, such as items on addition of matrices and the logic of matrix data storage for example which were equally suitable for both samples. Items introducing matrix multiplication were included in both tests but it was realised that their functions would be different for the two samples. The children who attempted test M were on the whole more able than those who attempted test A; many schools only introduce multiplication to the more able classes.

Matrices for data storage

The introductory question on both versions of the test used 3 × 3 matrices to store data. The items involved interpretation of the matrices, insertion of data into an incomplete matrix and addition. The facilities were high and for the overall sample (Test M and Test A) were over 80 per cent for those aged 14 and over 90 per cent for those aged 15. The data to be inserted in the incomplete matrix were deliberately not presented in the correct order and 2 per cent of the M sample and 6 per cent of the A sample did not re-arrange the data before entering it. On interview it was found that this tended to occur because the reading of the question was not thorough: mathematics questions are usually carefully worded and suffer from this careless type of reading. The addition question required the completion of an incomplete matrix, the children were told what it represented but not that the operation required to complete it was addition of the two previous matrices. On interview some less able children searched for number patterns to account for the given first row of the sum matrix and it was apparent that this number pattern approach was a strong draw for some of the less able children. The high facility levels of the items involved indicate that such a data storage approach provides an easily understood introduction to matrices.

The second question made the transition from matrices storing data to abstract matrices which had to be added, and examined the commutativity of matrix addition. The addition items produced high facilities (over 80 per cent for all items) but the commutativity question produced an interesting result. Having obtained two identical answers for two pairs of matrices under addition (A + B, B + A), the next question involved the realisation that the answers were the same – over 80 per cent of the total sample noticed this. When asked why this was so the rather brighter M sample performed much better, over 60 per cent answering correctly either that matrix addition is commutative or that when adding changing the order makes no difference. Of the A sample only 12 per cent of 14 year olds and 28 per cent of 15 year olds answered correctly. This may well be due to the commutativity of matrix addition only being stressed by teachers when comparing addition to the noncommutative operation of matrix multiplication. The A sample had not been introduced to matrix multiplication. Very few of the correct answers from the A sample mentioned the word 'commutative' while nearly two-thirds of the correct answers from the M sample did so. Approximately 10 per cent of the M sample gave an example of the addition of integers being commutative as part of their explanation of matrix addition being commutative. The children were told that the matrices had been swapped over but 20 per cent of the M sample and 40 per cent of the A sample attempted to restate this fact as their answer and, as on interview, showed either an inability to realise that swapping the matrices over might make the answer different or an inability to make use of the fact that different but related questions produced the same answer.

The necessary logic involved in successful matrix data storage was also examined and the two particular aspects looked at were the necessity of consistent positioning when storing data in matrix form and the need to use the same kinds of measurement. The information given was:

> Roger's stool needed 5 metres of canvas, 3 metres of wood and 4 metres of string. Simon's stool needed 4 metres of wood, 3 metres of canvas and 2 metres of string.

Item 24 involved finding what was wrong with the matrix

$$\begin{array}{c} \\ R \\ S \end{array}\begin{array}{ccc} C & W & S \\ \left(\begin{array}{ccc} 5 & 3 & 4 \\ 4 & 3 & 2 \end{array}\right) \end{array}$$

Item 25 involved finding what was wrong with the matrix below given:

Roger's new stool needed 1 metre of canvas, 50 cm of wood and 3 metres of string. Simon's new stool needed 1 metre of canvas, 1 metre of wood and 2 metres of string.

$$\begin{array}{c} \\ R \\ S \end{array}\begin{array}{ccc} C & W & S \\ \left(\begin{array}{ccc} 1 & 50 & 3 \\ 1 & 1 & 2 \end{array}\right) \end{array}$$

Table 11.10   The logic of matrix data storage

		Item 24	Item 25
M	14 yrs	69.4	81.2
SAMPLE	15 yrs	73.2	84.1
A	14 yrs	48.1	46.7
SAMPLE	15 yrs	50.0	62.3
OVERALL	14 yrs	60.0	65.9
	15 yrs	67.4	78.6

In general item 25 proved to be easier, the 50 cm stands out as a likely suspect, but this is unavoidable in the metric system. If feet and yards had been used as the mixed units it seems likely that the item would have proved rather more demanding. Item 24 was found to be rather more difficult and the problem may again be skimmed reading, but on interview this did not account for the inability of bright subjects to find an error. It may well be that children are used to finding such data helpfully given in the correct order and have not been forced to think about the consistency which is essential if data are to be sensibly stored in matrix form.

Matrices and networks

The first three items looked at the understanding of multiplicative route structure, which is basic if this method is used to introduce matrix multiplication. They were included in both versions of the test. Two further items involving the multiplication of matrices describing networks were only included in Matrices M.

Item 19

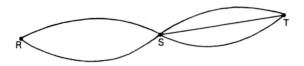

The diagram above was used for the first item. Children were told there are two roads direct from R to S and 3 from S to T and were asked to explain why there are 6 routes from R to T going through S. 'Label or number the roads if it will help you' was also added after experience on interview.

Item 20 asked for the number of ways from R to T after a new road is built from R to S (giving 3 × 3 = 9) and item 21 asked for the number of routes from "L to N" if there are 6 from L to M and 5 from M to N.

Table 11.11   Multiplicative route structure (percentage)

	Age	Item 19	Item 20	Item 21
M	14	75.9	77.1	69.4
SAMPLE	15	81.6	84.1	75.1
A	14	54.8	54.1	33.3
SAMPLE	15	56.6	60.6	45.9
OVERALL	14	66.6	66.9	53.4
	15	75.3	78.2	67.8

The incorrect answers are of interest, the essential structure is multiplicative but the draw of additive strategies remains strong and two types were quite common: straight addition and a more sophisticated strategy of 'add and add 1 more'. Table 11.12 shows the incidence of the addition strategies, as the item facility dropped the incidence increased. It can be seen that a significant number of successful answers were checked by diagram or worked out partly by diagram. Such children were clearly unhappy to rely on a simple multiplication to give the answer.

Further items included only on the M version of the test involved providing the correct row and column headings for a 2 × 2 route product matrix. After a careful

Table 11.12   Non-multiplicative strategies

Strategy:	Addition		Addition + 1		Correct and drew diagram
Items:	20	21	20	21	21
M 14 yr	4.7	11.2	5.3	1.8	11.8
M 15 yr	0.5	4.4	3.6	1.6	16.2
A 14 yr	6.7	12.6	4.4	5.9	5.9
A 15 yr	11.5	9.8	1.6	2.5	12.3

lead-in involving a simpler example less than half the M sample were able to provide the correct row and column headings.

The introduction of matrix multiplication by data storage: the athletics match

A rather long question examined the introduction of matrix multiplication using data storage. Matrices were used to store the number of 1st, 2nd and 3rd places achieved at a sports meeting and a column matrix was used to give the scoring system. Thus the product matrices gave the total number of points gained by a school or schools.

The early items involved the interpretation of the matrices involved and the formation of storage matrices (non-product) and produced high facilities, around 90 per cent for the M sample and 70 per cent for the A sample. The first product matrix involved the multiplication:

$$(4 \quad 1 \quad 3)\begin{pmatrix} 3 \\ 2 \\ 1 \end{pmatrix} = (\quad) \qquad \text{Success} \quad 90 \text{ per cent (M)} \quad 63 \text{ per cent (A)}$$

and could be worked out either as a straight matrix multiplication or from first principles (4 first places × 3 pts + 1 second place × 2 pts + 3 third places × 1 pt).

The second multiplication children were asked to perform was rather more complex and involved a 2 × 3 matrix:

$$\begin{pmatrix} 2 & 5 & 2 \\ 4 & 1 & 4 \end{pmatrix}\begin{pmatrix} 3 \\ 2 \\ \end{pmatrix} = \begin{pmatrix} 18 \\ \end{pmatrix}$$

Table 11.13   Completing and interpreting a matrix multiplication (percentage)

| | M Sample | | A Sample | |
	14 yrs	15 yrs	14 yrs	15 yrs
Multiplication correct	92		50	58
Intepretation	85		35	49

The most frequent incorrect answer was 'points' rather than an accurate description such as 'Cradley School's score'. In the interpretative items as a whole the most common finding was the vagueness of the answers.

The final items of the question for the A sample involved comparing the results of the multiplication above with the results of a similar multiplication using a different scoring system, (4 2 1) as a column vector. Over 40 per cent of the 14-year-old A sample and over 50 per cent of the 15-year-old A sample were able to complete the multiplications and note the comparison of the scoring systems. This supports the use of this method in introducing matrix multiplication.

The final items for the M sample involved completing the multiplication

$$\begin{pmatrix} 2 & 5 & 2 \\ 4 & 1 & 4 \end{pmatrix} \begin{pmatrix} 3 & 4 & 4 \\ 2 & 2 & 3 \\ 1 & 1 & 1 \end{pmatrix} = \begin{pmatrix} 18 & 20 & \\ & & 23 \end{pmatrix}$$

thus comparing three scoring systems for the same athletics match. Facilities for filling in the three missing elements were over 80 per cent for both the M samples. The previous item asking for an interpretation of the 23 in the final matrix produced a much lower facility, 52 per cent for each year group. Part of the failure rate was accounted for by vague answers failing to identify both the school concerned and the scoring system but 12 per cent of each year group answered '3rd places' indicating a lack of understanding of the results of the multiplication.

Interpretation of matrix products

Two questions were used to look at the ability to interpret matrix products, the first of which used the matrix multiplication shown below.

$$\begin{array}{cc} & \text{r} \quad \text{w} \\ \begin{array}{c} B \\ C \end{array} & \begin{pmatrix} 5 & 4 \\ 3 & 6 \end{pmatrix} \begin{pmatrix} 4 \\ 2 \end{pmatrix} = \begin{pmatrix} 28 \\ 24 \end{pmatrix} \end{array}$$

The square matrix gave information about the number of bottles of red wine and white wine sold by two wine shops, Browns and Carvers. Children were also told that red wine cost £4 a bottle and white wine £2 a bottle and were asked to interpret the numbers 28 and 24. The most common error involved ignoring the fact that the number of bottles has been multiplied by a price and will give an amount of money, not a number of bottles.

Table 11.14 Matrix product – interpretation (percentage)

Age	Correct		No. of bottles answer
14 yrs	74(M)	39(A)	17(M)
15 yrs	80(M)	46(A)	15(M)

The M test also contained two follow up items using the transpose of the square matrix above and asking for interpretations of the product matrix.

$$\begin{array}{cc} & B \quad C \\ \begin{array}{c} r \\ w \end{array} & \begin{pmatrix} 5 & 3 \\ 4 & 6 \end{pmatrix} \begin{pmatrix} 4 \\ 2 \end{pmatrix} = \begin{pmatrix} 26 \\ 28 \end{pmatrix} \end{array}$$

It is extremely tempting to say that the '26' represents the total cost of the red wine sold (9 per cent − 14 yrs, 7 per cent − 15 yrs), but this is not the case, the bottles of red wine sold at Carvers have been sold for the price of the white wine. In fact no rearrangement will produce a 2 × 1 matrix giving the total cost of the red wine sold and the total cost of the white wine sold.

A lead-in item asking how the 26 is obtained was answered satisfactorily by over 70 per cent of both year groups but only 42 per cent aged 14 years and 48 per cent aged 15 years could explain why the 26 was not the total cost of the red wine sold. The real problem appears to be that although many children are proficient at the mechanical act of multiplying two such matrices their understanding of the relationship between what the elements represent is often limited.

The second question looked at the problem of the construction of a matrix multiplication from given data. A 2 × 2 matrix was given, with accompanying information about its contents. The matrix gave the details of the number of buses and coaches used by two factories to bring people to work. The cost, per day, of running a bus and running a coach was given in the text and from this the children were asked to work out the cost per day for each factory. This involved correctly intepreting the matrix and multiplying by the appropriate cost (success rate: 85 per cent (M), 50 per cent (A)).

The M sample were then asked to write the whole 'sum' out as a matrix multiplication and were given the 2 × 2 matrix already positioned. Only 57 per cent of 14 yr olds and 63 per cent of 15 yr olds could do so (and this includes those who transposed the answer matrix). There appear to be problems when children are asked to construct matrix multiplications from a given story even when the order of the matrices concerned is moderate. Once again the relationships between the elements of matrices being multiplied are obviously not well understood.

Writing a story for a given matrix multiplication (M sample only)

Children were given the matrix multiplication $\begin{pmatrix} 4 & 7 \\ 3 & 8 \end{pmatrix} \begin{pmatrix} 1 \\ 2 \end{pmatrix} = \begin{pmatrix} 18 \\ 19 \end{pmatrix}$

and were asked to write a 'story' for it and asked what each of the 8 elements involved represented. Three scores were used to record the answers:
  1) form of necessary two-way classification table, e.g. two items at two shops
  2) relation of 1 and 2 to correct elements in square matrix
  3) sensible story e.g. 18 and 19 sensible entities.

Table 11.15   M sample (percentage)

Age	Correct classification	Correct structure	Satisfactory story	Vague multipliers	Wrong relationship
14 yrs	58	46	45	9.4	6.5
15 yrs	53	39	34	13.2	7.1

The superiority of the younger children on this question remains to be explained. Less than 1 per cent of stories were copies of other questions on the test or modifications of them. The majority of stories used prices for the numbers 1 and 2 and many were set in the school situation.

Writing a story for a given matrix addition (A sample only)

The question, write a story about the addition of two matrices to give a third matrix, was written along similar lines to the multiplication story question for the M sample. The addition structure allows more choice for the structure of each matrix, but all three matrices must have the same structure, like must be added to like. The ideal structure is still $\begin{pmatrix} A_1 & B_1 \\ A_2 & B_2 \end{pmatrix}$ which makes full use of the data storage capabilities of matrices but sensible stories can be constructed using a structure involving four separate entities generalisable to $\begin{pmatrix} A & B \\ C & D \end{pmatrix}$ which was far more popular, being chosen by over one third of the sample. The main incorrect strategy involved not keeping the structure of each matrix consistent and hence not adding like to like. Twenty-five per cent of the 14 yr olds and 31 per cent of the 15 yr olds gave adequate stories and it is interesting to note that over 80 per cent of the acceptable stories indicated that the third matrix gave the sum of the other two, i.e. represented the total of what was being added. This explicit use of the word 'total' was the greatest difference between successful and unsuccessful stories.

Matrix multiplication and commutativity (M sample only)

The question gave two complete matrix multiplications, of form A x B and B x A, and asked for a conclusion to be drawn after comparing the multiplications – the answers were given, as this removed the possibility of calculation errors distorting the answers to the question. Sixty per cent of each year concluded correctly that matrix multiplication is not commutative. About one third of the correct answers mentioned the word commutative while two-thirds argued along the lines that when multiplying matrices a change of order is important. It is interesting to note that 3 per cent of all answers involved the wrong 'jargon', using either associative or distributive instead of commutative. On interview several subjects were unable to draw any conclusions even when helped and prompted. The concept of commutativity clearly does not form part of the mathematical equipment of many children.

Transformations (M sample only)

The matrices used were $\begin{pmatrix} 1 & 0 \\ 0 & -1 \end{pmatrix}$ and $\begin{pmatrix} 0 & -1 \\ 1 & 0 \end{pmatrix}$ and a right angled triangle was transformed. Facilities for calculating the new positions of the vertices were over

67 per cent for each year group. Facilities for drawing the triangle in its new position were rather lower and few subjects who made one error in calculating a new vertex seemed able to recognise the error from the shape of the new (distorted) triangle.

The most interesting errors observed resulted from noticing that $\begin{pmatrix} 1 & 0 \\ 0 & -1 \end{pmatrix}$ changed the sign of the second coordinate of the figure and assuming that $\begin{pmatrix} 0 & -1 \\ 1 & 0 \end{pmatrix}$ would merely change the sign of the first coordinate. On interview few children who took this view bothered to check by calculating one new position to confirm their hypothesis.

The hierarchy

Of the 33 items common to both A and M test papers, 27 were used to form a Core Hierarchy. The facilities given below are for the A and M samples combined. General descriptions of the items are given.

Group 1: Addition and simple interpretation
(i) The addition of matrices within a 'story' context or with no meaning attached, (ii) the interpretation of matrices given either row and column headings or the information in written form, (iii) writing data in matrix form, (iv) the extraction of data from a matrix and simple calculation on such data.

Group 2: Further interpretation and work with elementary product matrices
(i) The interpretation of simple column matrices (no headings), (ii) the explanation of a simple matrix multiplication (data given in the text), (iii) the calculation of a single entry product matrix and the comparison of this with another single entry product matrix (data in the text), (iv) the calculation of one entry of a 2 × 1 product matrix when the other has been calculated and defined.

Group 3: Interpretation of elementary product matrices and elementary route structure
(i) The identification and calculation of a number of routes from a given diagram where the number is small (3 × 3 = 9), (ii) the significance of the consistency of units of measurement in a data storage matrix, (iii) the interpretation of one element of a 2 × 1 product matrix when the other element has been calculated and defined, and the calculation of such an element when the other has not been defined, (iv) the extraction of information from a storage matrix and the performance of calculations on this to give elements of the product matrix.

Group 4: Advanced interpretation, multiplicative route structure and the commutativity of addition
(i) The calculation of the number of routes in a route system (no diagram) when the number is large — 30, (ii) the recognition of errors of position in a data storage matrix, (iii) the interpretation of one or both elements of a 2 × 1 product matrix, (iv) the recognition that the addition of matrices is commutative.

Table 11.16   Comparison of year groups (percentage of relevant sample)

Level	Overall Facility (pass mark)	14 yr	15 yr	M Sample 14 yr	M Sample 15 yr	A Sample 14 yr	A Sample 15 yr
0		7.5	5.5	1.0	0.5	15.5	20.5
1	88–95 (5/8)	10.0	2.5	2.5	0.5	20.0	8.0
2	80–84 (3/5)	14.5	9.0	10.0	6.5	20.0	17.0
3	71–75 (5/8)	16.0	15.0	14.5	14.5	18.0	16.5
4	49–67 (4/6)	52.0	68.0	72.0	78.0	26.5	38.0

Note: 5.7 per cent (non-scale types) have been recoded and included.

Extra items contained in the M test and attempted by the M (brighter) sample enabled the hierarchy to be extended in terms of items with greater complexity, these are given as groups 5 and 6.

Group 5: Commutativity and complex classification

7 items   Pass Mark 5/7   Facility Range 65 to 52 per cent (M Sample) (i) The distinction between the non commutativity of multiplication and the commutativity of addition, (ii) the drawing of transformed geometric figures, (iii) the accurate 2-way classification of a complex product matrix, (iv) the writing of a suitable story as a matrix multiplication, (v) the giving of the correct structure to a matrix so that matrix multiplication can be placed in a story context.

Group 6: The relationship between the elements of a matrix multiplication

4 items   Pass mark 3/4   Facility Range 46 to 38 per cent (M sample) (i) The use of the correct row and column headings on a route product matrix, (ii) the understanding of the relationships between the elements of matrices being multiplied (including writing a story for a matrix multiplication).

Commentary and implications for teaching matrices

The hierarchy contains two central themes, the ability to interpret increasingly complex matrix forms and the ability to select and use an appropriate matrix form for a given situation. The two are naturally related by an increase in the ability to understand the relationships involved. In the first case the relationship between the rows and columns of a particular matrix is important and the recognition of the relationships between the elements of matrices being multiplied is an important part of both.

The high facilities of the items involving data storage (both as matrices and in matrix multiplication) show that this approach appears to be easy for the child. Even half of the A sample, after a brief introduction could deal with matrix multiplication in this form. The data storage method of introducing matrix multiplication places this somewhat arbitrary operation in a concrete setting and provides a rationale for it. Common sense may be applied by the child to advantage.

The two items examining the logic of data storage indicated that errors in such matrices are not easy to spot. Inconsistent units of measurement make just as much of a nonsense of them as transposed or misplaced entries and while the latter fault is sometimes given attention in text books the former is often not stressed. It seems that examples on data storage where the order of presentation of data is varied might well be useful.

The introduction of matrix multiplication by route matrices appears to be less justifiable. The inherently multiplicative structure of routes via an intermediate point does not come easily, and the additive strategies sometimes used to replace the correct multiplicative one obviously have a strong attraction. Most other methods of introducing matrix multiplication do not lend themselves to the attachment of meanings to the individual elements. It is difficult to attach meanings to the elements of a transformation or decoding matrix and thus provide the concrete basis for matrix multiplication which appears to be helpful.

The general facility level of the items involving matrix multiplication suggests that the topic is within the grasp of the majority of third and fourth year pupils, but the results indicate that the understanding is often rather mechanical. Greater emphasis on the relationship between the elements of matrices being multiplied seems likely to produce a less mechanistic view of matrix multiplication. The multiplicative relationship in the situation being modelled must be mirrored by the relationship between the elements of the matrices involved.

Even the simple item on the commutativity of matrix multiplication proved to be demanding and many children did not see the presence or absence of commutativity as notable. Structure in general appears to be a difficult subject for children and the inclusion of matrices in a syllabus mainly to demonstrate structure is a debatable strategy.

12 Comparison of performance between year groups

Every test paper was attempted by children from several age groups (usually three) so that a comparison of performance on the same items could be made. In preceding chapters item facilities and the number of children at each level in the specific topic hierarchy have been given for each age group tested. The gaps between levels in a hierarchy were very seldom equal and it was often the case that items in say Level 3 were much more difficult than those in Level 2 whereas the facilities of the hardest Level 1 items were very close to those of the easiest Level 2 items. In particular topics a child who obtained a $\frac{2}{3}$ pass mark on Level 3 items displayed a grasp of a type of mathematics different from that represented by items in Level 2, e.g. in Graphs, Level 3 items were concerned with algebraic relations while Level 2 items involved interpretation of data graphs. There is no suggestion that the step from Level 2 to Level 3 is equivalent to that from Level 1 to 2 or indeed that the steps between any two levels can be regarded as equal. It is only in the context of the topic chapter and the facility ranges of each level that the reader can make a decision about the relative gains a child has made when he attains the pass mark on a set of items. Because the items at a particular level were associated and because those same items were usually in problem form we have referred to the levels as 'levels of understanding'. They are of course measures of attainment on particular tests but since their formation required rather more than total scores and the attainment of each infers the attainment of all easier levels, the claim is made that they to some degree represent the child's grasp of the topic.

Individual topics

The general picture that emerges when one looks at the performance of say 13 year olds and compares this with success rate of older children, is that there is an increase of five to ten per cent in the facility of items over two years. In some topics this pattern is not apparent. For example in Algebra there is a large gap between the performance of the 13 year olds and the 14 year age group, the facilities of items for the 14- and 15-year-old samples are rather closer together. The 13 year olds will have had little experience of generalised arithmetic in their mathematics lessons and the distinction between their results and that of older children is probably due to this. The distinction between year groups in Measurement appears to relate to the ability to use a formula for area and volume problems. The progress year to year is still of the order of five to ten per cent however. In Graphs, the 14 year olds were not significantly better than the children of 13, the third year in school seemingly being a consolidation period for this topic. Similarly the distinction between the 15 year olds and younger groups in the topic of Positive and Negative Numbers was small, and on some items the younger children gave a considerably better performance. This may be due to a desire on the part of the older children to rationalise the rules of operation, e.g. minus times a minus makes a plus, the simple acceptance of which was sufficient for the younger age group.

In Fractions the older children were always better at solving problems but the 12 year olds showed considerable competency in addition computations.

Longitudinal survey

The picture presented so far is that of a comparison between sets of different children whose ages differ by one year. Further information is available from a longitudinal study carried out over two years (three testings) in the topics of Algebra, Ratio and Graphs. For each topic 200 children who were in their second year of secondary education in 1976 and whose IQ score was known were tested during July in three consecutive years. Each group of 200 was selected from four schools and randomly within four IQ ranges − IQ ≤ 89, 90 ≤ IQ ≤ 99, 100 ≤ IQ ≤ 109, IQ ≥ 110. By the summer of 1978 the number of children who had been tested three times had dropped to about 100 in each topic (absence and change of school being responsible for the drop out). It is possibly worth noting that there were many more absences within the lowest IQ band than within the other three.

The results of the three testings in each topic are described below. It should be borne in mind that the research team had no information on the child's performance in the classroom, on other areas of mathematics nor indeed on any aspect of learning except performance on the tests. The word 'progress' is used in this context and means the attainment of the $\frac{2}{3}$ pass mark on a level higher than that previously passed.

Ratio

The performance on the Ratio test appeared to be closely related to the IQ score of the child. This was apparent from the start when no child of IQ score less than 89 (standardised score) achieved Levels 3 or 4 and only two were at Level 2. In the group with IQ score greater than 110, no 13 year old was at Level 0 although many were at Level 1. Figure 12.1 shows the percentage of children at each level in 1976 and 1978.

As can be seen there is progress in each IQ group but for the lowest IQ group it does not go beyond Level 3 while for those with the highest IQ score the number who are at Level 4 has quadrupled in two years. Note however that not all of those with IQ score greater than 110 had achieved the $\frac{2}{3}$ pass mark in Level 4 by the age of 15. In the IQ score range 100 ≤ IQ ≤ 109 the children were dispersed equally between Levels 1, 2, 3 with 17 per cent at Level 4.

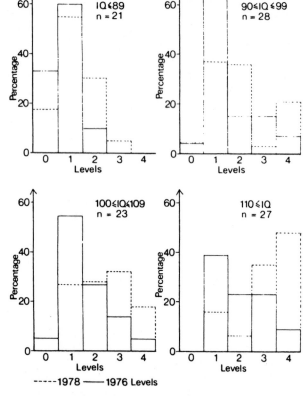

Fig. 12.1   Progress over two years in four IQ groups (Ratio).

Progress of individual children in Ratio

The pattern of movement level to level for individual children over the two years is shown in Table 12.1.

Table 12.1   Progress of individual children in Ratio

Moved	Total n = 99	IQ ⩽ 89 n = 21	90 ⩽ IQ ⩽ 99 n = 28	100 ⩽ IQ ⩽ 109 n = 23	IQ ⩾ 110 n = 27
0 levels	41	12	13 (1 at L4)	6	10 (4 at L4)
1 level	26	5	6 (1 to L4)	9	6
2 levels	14	2	4 (3 to L4)	2	6
3 levels	2	0	0	1	1
Regressed	6	1	3	2	0

Ten non-scale types in 1978 omitted
Note: a regression is a failure in 1978 to achieve the pass mark of $\frac{2}{3}$ on a group of items which had been passed in 1976.

It should be noted that some children regressed at the second testing (1977) and then returned to or progressed beyond the position they held previously. Others were non scale types (attaining the pass mark at a higher level without achieving it at all lower levels), whilst some failed to pass a set of items but obtained a better mark at the next hardest level. For example, in the IQ score range $90 \leqslant IQ \leqslant 99$ nine children apparently regressed to a lower level at either the second or third testing. Five of these however were non scale types on one or other of the tests. Three improved their marks on the easier groups of items or on the set immediately after that which they now failed (having succeeded on it earlier). One regressed in the second year and then improved in the third year and one regressed in the second year and stayed at that level.

Most of the fluctuations in levels occurred at the second testing rather than the third. The diagram (Fig. 12.2) shows the percentage of children at each level after two years and where those same children were in 1976 when they were aged 13.

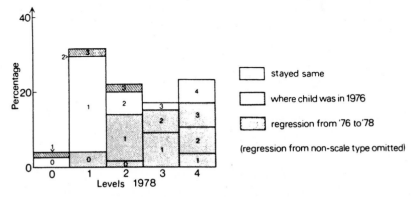

Fig. 12.2   Improvements and regressions, Ratio (n = 99).

The progress is slow and many children reach only Level 2 at age 15. This implies that children can on the whole cope with doubling and halving and items in which they can build up to an answer by repeating an amount but any further complexity defeats many.

Addition strategy

The error made by many children when faced with a Level 3 or 4 Ratio question was to attempt a solution by adding the difference a − b rather than using the ratio a/b. This particular strategy was fairly persistent in that 37 children in the longitudinal sample were using it consistently in 1976 and 16 of these were still using it in 1978. Six children were now successfully dealing with the four items on which a large number of children had used the strategy, the remainder were still failing to solve them but using the addition strategy on two and another incorrect method on the others.

Graphs

The number of children with IQ less than 89 who did the Graphs test in 1976 had by 1978 been reduced by absence on one or more testing from 13 to 2. The results of this group are therefore included with the next highest IQ group in the following Figure.

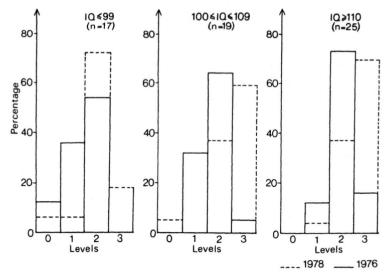

Fig. 12.3   Progression by year in Graphs.

The increase in the number of children reaching Level 3 (the highest level) is most marked for the two higher IQ groups. The facilities for Level 3 items for the general 13-year-old sample are all between 20 and 7 per cent, for the 15 year olds the same items have facilities between 37 and 14 per cent. There is therefore an obvious improvement in attainment over the two years. Recall that the general sample was chosen to represent the *normal* distribution of IQ and the longitudinal sample described here was selected to represent IQ groups equally. The longitudinal group with IQ score less than 99 showed some improvement but less than 20 per cent reached Level 3. Two thirds of those at Level 2 scored less than $\frac{3}{11}$ on Level 3 items. Of those children who made no apparent progress level to level over the two years 11 stayed at Level 2. On the other hand in 1978, 30 per cent of the group, $100 \leqslant IQ \leqslant 109$, obtained full marks on the set of items in Level 3 while half of those in the top IQ group did so. The movement of individual children is shown in Table 12.2.

Table 12.2   Progress of individual children in Graphs

Moved	Total n = 61	IQ ⩽ 99 n = 17	100 ⩽ IQ ⩽ 109 n = 19	IQ ⩾ 110 n = 25
0 levels	22	8	4 (1 at L3)	10 (4 at L3)
1 level	36	8 (3 to L3)	13 (9 to L3)	15 (13 to L3)
2 levels	1	1	0	0
Regressed	1	0	1	0

One non scale type in 1978 omitted.

Note that with only three levels in the Graphs hierarchy not a great deal of movement was possible. More children reached Level 3 than might have been expected from the picture obtained from the wide scale testing.

Algebra

The analysis of the test on generalised arithmetic resulted in four levels of understanding. There was a distinct difference in attainment (by the age of 15) between those with an IQ score below 100 and those with IQ score above 100. There were no children of IQ score greater than 100 who were still at Level 1 whereas 40 per cent and 17 per cent of those in the IQ ranges IQ $\leqslant$ 89 and 90 $\leqslant$ IQ $\leqslant$ 99 respectively were still only achieving this level, see Fig. 12.4.

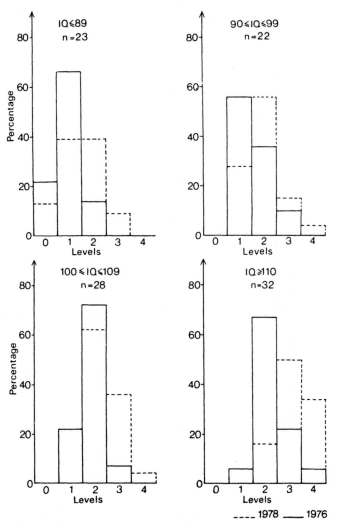

Fig. 12.4   Progression over two years in Algebra.

The progress to Level 4 where the first true use of variables occurs, was most marked for the highest IQ group, this level of abstraction being beyond the attainment of the lowest IQ group. There was some progress in every group however over the two years, although the pattern level to level differed with each IQ range. Nine children (out of 105) moved two levels in two years, five of these had an IQ score of more than 110. The details of movement for individual children is shown in Table 12.3.

Table 12.3   Progress of individual children in Algebra

Moved	Total n = 105	IQ ≤ 89 n = 23	90 ≤ IQ ≤ 99 n = 22	100 ≤ IQ ≤ 109 n = 28	IQ ≥ 110 n = 32
0 levels	42	9	14	13	6
1 level	51	10	7 (1 to L4)	14	20 (6 to L4)
2 levels	9	2	1	1 (to L4)	5 (4 to L4)
Regressed	3	2	0	0	1

Summary

The results of the longitudinal study confirm the description of a slow progression of attainment year to year obtained from the topic hierarchies (based on data from wide-scale testing). The percentage of children at each level in a hierarchy was calculated on the attainment of the $\frac{2}{3}$ pass mark and no allowance was made for a lesser mark in a higher group, so children may have achieved a greater degree of success than is apparent. The inability to take the step from one level to another is sometimes due to the large facility gap between two levels (e.g. Graphs) but in other cases it may point to the need for considerable consolidation and extra experience before a child can make a crucial step.

The performance for each sample in the longitudinal study is closely related to measured IQ. There is a progression for the lower IQ groups (IQ ≤ 100) but since many of these children are at Level 0 (unable to make a coherent attempt at even the easiest items) when aged 13, their progress does not go beyond Levels 2 or 3. It is demanding a great deal to expect them to progress faster than the children with higher IQ score within the same time period. We have no data on 16 year olds so whether given more time most children would progress to Level 4 is open to question. Indeed since most of those not taking public examinations in mathematics leave school at age 16, it is unlikely that a true picture of their capabilities will ever be available.

In the next chapter the difficulty of matching hierarchies is discussed and an attempt is made to give a guide to the comparable difficulty of different areas of the secondary school mathematics curriculum. The data used are those obtained from the wide scale testing. The results of the longitudinal study quoted in this chapter support the use of such data. Bearing in mind the composition of the two samples there is no glaring mismatch between the results. The very few regressions found in the longitudinal data (admittedly over two years rather than one) add weight to the argument that the results of the larger survey can be regarded as coming from the same children tested in consecutive years.

13 The hierarchies

Each chapter in this book has been devoted to the description of a mathematics topic commonly taught in English secondary schools. The description has included the methods used by children to solve problems, the errors they make and a hierarchy of levels of understanding based on the analysis of the results of wide-scale testing. In this chapter an attempt is made to match the hierarchies against each other in order to provide a total picture. A match between topic levels should enable a teacher to have an approximate guide to the comparative difficulty of topics.

An exact match between topics is not possible for two reasons: (i) the number of levels in each topic hierarchy varies; (ii) the facility range of items within a level varies topic to topic. For example when one tries to match other topic hierarchies against that of Graphs one finds that two levels of Positive and Negative Numbers match Level 1 Graphs in facility and no Ratio level corresponds to Level 2 Graphs.

The general picture topic to topic was obtained by 1) looking at the performance on two tests of certain children and 2) matching hierarchies based on the results of testing one group of children against the results of a second group who did a different test (the two samples were the same age and were representative of the normal distribution on I.Q. score).

The number of children in a subsample who did two papers was sometimes small but a comparison of performance was made for each pair of tests. For example the cross tabulation of performance on the Ratio and Graphs tests in 1976 (n = 439) is shown overleaf. Although it can be seen that a clear matching of levels is not apparent, this is not surprising since for example there is no Graphs level of the same facility range as Level 2 Ratio, nevertheless it is clear that children below Level 2 on Ratio do not generally perform at Level 3 on Graphs and vice versa.

A statistical analysis of the correspondence between achievement on any two tests was carried out using the γ coefficient which quantifies the probability that if a child A is better than child B on one test he is also better than child B on another test. The γ coefficients obtained varied from .85 to .59, being mostly nearer .8 (maximum $\gamma = 1$) and are listed in Appendix 2. These values of γ calculated on the performance of the same children on two tests support the hypothesis that there is no gross mismatch between achievement in two mathematical areas and so the results from the general sample (10 000 children) were used to provide a matching of hierarchies.

Having accepted that no exact match was possible an approximate guide to the comparable difficulty of levels in each topic was obtained by taking four sections of

Table 13.1   A comparison between performance on Graphs and Ratio
(Age 13, 14, 15)

		Graphs Levels					
		0	1	2	3	ROW TOTAL	
	0	13	13	2	0	28	Cell entries are
	1	29	105	73	2	209	numbers of children
Ratio levels	2	3	21	78	12	114	
	3	0	4	37	11	52	
	4	0	0	20	16	36	
COLUMN TOTAL		45	143	210	41	439	

the facility range 0–100 per cent. Each section contained a level or levels from topics, the selection being made on the facility range of those levels, so that for example the easiest section would contain Level 1 from most topics. In addition a scrutiny of the types of items which occurred within a section was made in order to ascertain any greater degree of complexity in one topic than another. For convenience the four sections are called Stages.

The test papers were usually given to three age groups, although these might vary; for example ages 12–14 for Measurement and 13–15 for Algebra. All the test papers except Number Operations were attempted by the 14-year-old sample so the stages are based on the results obtained from testing this age group and then applied to each of the other age groups. The contents of each stage are fixed for all age groups in that Stage 1 for the 13-year-old sample contains exactly the same items as Stage 1 for the 15 year olds but the hardest items (providing the lower limit of the stage) are solved by less of the younger children than by the older sample. In addition not every topic appears in every stage since some topics did not have items within certain facility ranges. The lower limits of each stage, i.e. the facilities of the hardest items within that stage, are shown in Table 13.2. The progress from year to year is not very great and has been described in the previous chapter.

Table 13.2   Limits of four stages – hardest items
(percentage)

	13 yrs	14 yrs	15 yrs
Stage 1	70	70	80
Stage 2	48	50	58
Stage 3	18	22	30
Stage 4	2	5	8

A general description of the type of item found in each stage together with examples from the appropriate topics follows. The levels which occur in each stage are given in order that the reader can refer to the appropriate chapter for more details. There are children of course, who did not achieve Level 1 on a topic and who have been designated Level 0. Although we are aware of the items on which these children fail we have no information on the type of item on which they might succeed.

Stage 1

This includes the easiest levels of most of the topics. There are some children, probably about 10 per cent, who cannot achieve a ⅔ criterion pass mark on Stage 1 items (see number of children labelled Level 0 in each of the topic chapters). It is not known what areas of knowledge these children possess. Each of the tests assumed that numbers up to 20 were known, numbers up to 1000 recognised and that the child had the ability to repeat a unit of measure in order to find the length or area of an object. All topics which are covered in the Junior school curriculum. Stage 1 items extend these concepts and provide an extra demand. The extra requirement might be a new notation or the use of one half as well as whole numbers. The fractions questions at this stage are of the type:

1.   Shade two thirds

or

2.   What fraction is shaded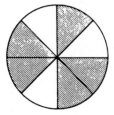

or

3.   In a bakers shop ⅜ of the flour is used for bread and ⅜ of the flour is used for cakes. What fraction of the flour has been used?

These items test the meaning of fractional notation and the first ideas of addition. The dimension added to the whole number knowledge is therefore the relationship between part and whole. The same relationship in Decimals extends the number system to include the meaning of the decimal point. To solve other Stage 1 decimal questions the child must also be able to compare whole numbers beyond those which can be merely counted.

The questions of comparable difficulty in Ratio and Proportion deal with the ratios 2:1 and 3:1. The child must recognise the need to halve, e.g. in the recipe question he must note that 8:4 people implies a halving relationship for the appropriate ingredients or he must realise there is a pattern to the increase in food in the eel question, e.g.

The eels are fed sprats, the number depending on their length. If C is fed two sprats, how many sprats should B and A be fed to match?

|—————————————— 15 cm ——————————————| A

|————— 10 cm —————| B

|—— 5 cm ——| C

Although the easiest Graphs questions are purely pictorial representation of data, some items involve the plotting of a point given two coordinates. Stage 1 Reflection items require either coordinates or the control of both distance and direction. On both the Graphs and Geometry papers the provision of a grid means children can count to obtain the positions of points but they must know that counting is appropriate. The measurement questions also require counting to find an area but the added dimension is that half squares are involved, not just whole square centimetres.

The Vector items (only 3rd and 4th years attempted this paper so we can draw no conclusions as to the possible degree of difficulty for 2nd years) involve new notation in the form of a column matrix and multiplication by a scalar, e.g. $2\begin{pmatrix} 1 \\ 2 \\ 3 \end{pmatrix}$. This is a use of whole numbers which certainly does not appear in the primary school. Similarly in positive and negative numbers the child can interpret $^-2$, probably as a shift along a number line and he can then add. Multiplication of intergers is also successfully carried out but probably solved by an application of a rote learned rule, i.e. 'Minus times a minus makes a plus' etc. In Algebra letters are collected and are probably viewed as objects but this is also one stage removed from collecting numbers and consequently appear as harder Stage 1 items.

The hardest items in Stage 1 tend to show complexities in the use of conventions, for example the plotting of $(1\frac{1}{2}, 4)$ in Graphs (there is about 15 per cent drop in facility from plotting whole number pairs). On the vector paper $4\begin{pmatrix} 4 \\ b \end{pmatrix} = \begin{pmatrix} a \\ 8 \end{pmatrix}$ is a hard Stage 1 item, here 'a' must be found and letters are used for the first time. In measurement the hardest item in Stage 1 is to find the volume of a cuboid $(1 \times 1 \times 3\frac{1}{2})$ which is still possible by counting but which involves three dimensions shown on a two dimensional diagram.

The conceptual demand of the Stage 1 items is greater than a knowledge of just whole numbers. The whole numbers must be viewed in pairs (coordinates), triples (vectors), as multipliers (ratio) or divisible (decimals). Many of the items are testing the knowledge of a new mathematical vocabulary e.g. tenths, halves, reflections. Children capable of solving Stage 1 items could be regarded as knowing 'meanings' in the mathematical language, but nearly thirty per cent of our child population (even at age 15) can go no further i.e. cannot apply this language.

Many of the Stage 1 items have a visual aspect, diagrams are given or grids provided and the child usually needs to perform only one step albeit with two elements for completion of an item. Counting is one such step and plotting a number pair is another.

In order that readers can compare Stage 1 with the individual topic hierarchies, Diagram 1 shows which levels of which topic are included in this stage. Diagram 2 gives a brief description with illustrative items of the contents of Stage 1.

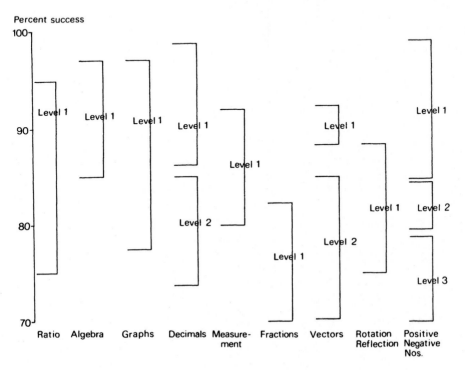

Diagram 1   Stage 1 levels   Facilities from 3rd year sample.

Diagram 2   Stage 1

Ratio	Fractions	Decimals	Vectors	Positive and negative nos
No rate given, × 2, × 3 need be derived from question context.	Representation of parts of a whole in concrete setting and 1st operation on fractions (Addition). Equivalence from doubling.	Place value with whole numbers up to thousands, decimal notation as far as tenths.	Addition . Scalar × . Simple equations. Realise need to use neg. ints in moves.	Directed numbers located on a number line. Addition using shifts. Multiplication by use of a 'rule'.
Recipe for 8 people contains 2 pints water How much water for 4 people? How much water for 6 people?		Write down a number between 4100 and 4200	Move of form $\binom{a}{b}$	Moving $+5$ then -4 gives . . . this can be written $\ldots + \ldots = \ldots$
				Put these numbers in order of size; start with -7
		This number is	Work out $2\binom{7}{5}$ $\binom{4}{}$	$-3, -7, +3, -5, +4$
	Labelling $1/3 = 2/?$	0.2 The 2 stands for 2 . . .	From C to D (given diagram) the move is described by vector $\binom{2}{-1}$	
			Why do we write -1?	

Graphs	Algebra	Measurement	Reflection and rotation
Easy plotting of points. Given scattergram or block graph read off specific points. Recognition that straight line corresponds to constant rate.	Require no knowledge of significance of letters (letter can be ignored, evaluated or seen as an object)	Measuring the area of rectangles and triangles by counting the number of square and half square units.	Can cope with two dimensions, (e.g. direction × distance), in step by step manner, but not critical of answer.

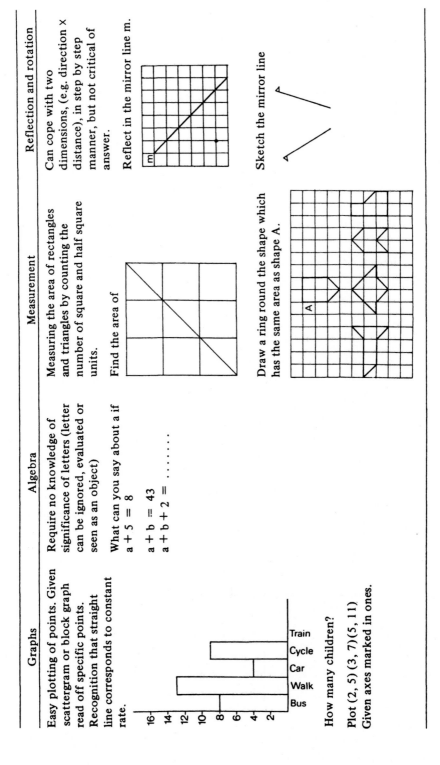

Algebra:

What can you say about a if

$a + 5 = 8$

$a + b = 43$
$a + b + 2 = \ldots\ldots\ldots$

Measurement:

Find the area of

Draw a ring round the shape which has the same area as shape A.

Reflection and rotation:

Reflect in the mirror line m.

Sketch the mirror line

Graphs:

How many children?

Plot (2, 5) (3, 7) (5, 11)
Given axes marked in ones.

Train
Cycle
Car
Walk
Bus

Stage 2

The position of the line drawn to delineate the end of Stage 2 meant that two levels of understanding in Decimals, two in Measurement, two in Matrices and two in Vectors were included, but none in Positive and Negative Numbers or Ratio. This was because there was a 20 per cent facility gap between Levels 1 and 2 on Ratio and also on Positive and Negative Numbers whereas on Decimals etc. the two levels tended to be close in facility. Stage 1 items were mainly concerned with understanding the meaning of new conventions, whereas Stage 2 seems to be concerned with the application of these conventions and is therefore more concerned with the understanding of when they should be applied than with the pure knowledge of the language or symbols. What was introduced in Stage 1 becomes operative in Stage 2 and there are possibly two dimensions of difficulty in the problems. In Graphs the child is no longer asked to plot a point but to interpret a point on a scattergram or a set of points on a time-distance graph for example:

7.   John leaves home to go to a disco in Cambridge 3 miles away. He walks one way and takes the bus the other way. This is a graph of his journey:

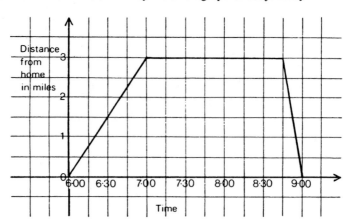

At what time did he get to the disco?
What do you think he was doing between 7.00 and 8.45?

In Measurement something more is needed than just counting squares when all are shown; when counting cubes to find the volume the child has to take into account hidden cubes and the counting is on a two dimensional diagram not a three dimensional block. The formula for the area of a rectangle is needed for the first time, e.g.

Shaded Area = . . .

16. The area of the striped rectangle A is 12 square centimetres. What is the area of the dotted rectangle B?

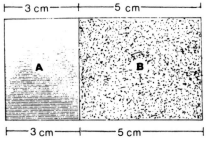

Area of dotted rectangle B: . . .

It is noticeable that when the area of a rectangle (10) is given together with one dimension (4) and the child is required to find the second dimension, the item is much harder than the second example just shown (50 per cent success against 76 per cent success for the third years). Both require more than simple multiplication but the amounts A = 10 and L = 4 of course result in the width being a fraction.

In Fractions the important aspect of equivalent fractions occurs at Stage 2, e.g.

$$\frac{2}{3} = \frac{?}{15} \text{ or}$$

A relay race is run in stages of ⅛ km each. Each runner runs one stage. How many runners would be required to run a total distance of ¾ km?

Although the second item might be regarded as an application of equivalent fractions the equivalence ¾ = ⁶⁄₈ is one that is very familiar. In Algebra though the letters are still used as objects or can be evaluated there is an increase in complexity and when items involve simple collection two letters are involved, e.g.

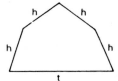

Perimeter = . . .

The collection which in Stage 1 resulted in 3p has now to be added to a second collection say 4q.

In Reflections a more critical understanding is required; for example children now have to cope without a grid (for the slanting mirror and single point) and be able to recognise that there is *no* mirror line in the example below. However, children are still not able to coordinate the slopes of an object and a slanting mirror line accurately.

The meaning of hundredths seen in their simple connection to tenths i.e. ten of one makes one of the other, is a Stage 2 decimal concept, for example:

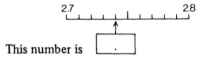

This number is [.]

0.18   0.1 / 10 / 0.2 / 20 / 0 / 1 / 2

Ring the number nearest in size to 0.18

The questions involving tenths require the child to convert from one decimal place to another, e.g. Add one tenth to 2.9, the obvious incorrect answer being 2.10 where the child does not make the connection between units and tenths. The easier Stage 2 Vector items involve the representation of a vector as a directed line segment and many of the items test the understanding of the implications of this definition, e.g. the interconnection between $\underline{c}$, $-\underline{c}$ and $2\underline{c}$. The idea of a vector as a column matrix was introduced in Stage 1, a second meaning is given in Stage 2.

Stage 2 extends and uses what was contained in Stage 1, in some cases the vocabulary of mathematics is still being built up, e.g. Decimals and Vectors, in other cases that vocabulary is being used. Problems may require two steps for completion or need an interpretation in the form of a graph or the application of formulae. The child can no longer simply carry out an instruction, he must be in a position to see what is needed for solution. Although he is not required to invent a problem solving strategy he has to ask himself 'what method do I use'. Diagrams 3 and 4 illustrate the contents of Stage 2.

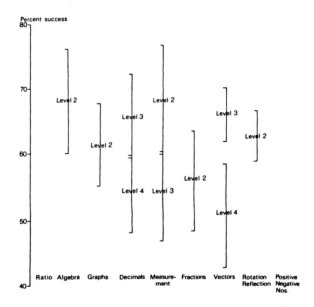

Diagram 3   Stage 2 levels   Facilities from 3rd year sample.

Diagram 4   Stage 2

Ratio	Fractions	Decimals	Vectors	Positive and negative nos
No items	Equivalent fractions not obtained by doubling. Using equivalence to name parts with familiar fractions or when diagram provided.	Hundredths, thousandths. Relationship between adjacent columns.	Free vectors. Identify $2\underline{a}$ etc. given $\underline{a}$. Elementary route vectors, Move of form $\begin{pmatrix} a \\ -b \end{pmatrix}$	No items
	Shade 1/6 of the dotted section of the disc. What fraction of the whole disc have you shaded?	Ring the number NEAREST IN SIZE to 0.18 0.1 / 10 / 0.2 / 20 / 0 / 1 / 2	Simple translations. Elementary abstract moves. Elementary equivalent vectors. Move of form $\begin{pmatrix} 0 \\ -b \end{pmatrix}$.	

A relay race is run in stages of $\frac{1}{8}$ km each. How many runners are needed for $\frac{3}{4}$ km?

2.7 —————|————— 2.8

This number is ▭ . ▭

if $\begin{array}{c}\nearrow\end{array}$ is $\underline{a}$, then $\begin{array}{c}\nearrow\end{array}$ is $2\underline{a}$ $\begin{array}{c}\nearrow\end{array}$ is $-\underline{a}$

What is $\begin{array}{c}\nearrow\end{array}$?

Write $\begin{pmatrix} 2 \\ 9 \end{pmatrix}$ using $\begin{pmatrix} 1 \\ 0 \end{pmatrix}$ and $\begin{pmatrix} 0 \\ 1 \end{pmatrix}$: $\begin{pmatrix} 2 \\ 9 \end{pmatrix} =$

Diagram 4   Stage 2 (cont'd)

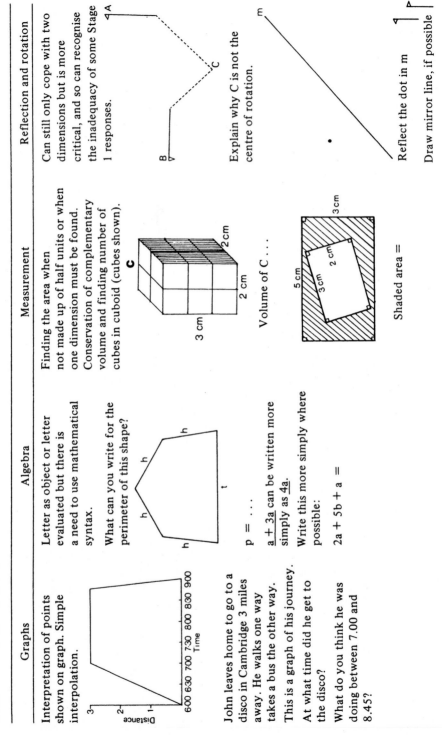

Graphs	Algebra	Measurement	Reflection and rotation
Interpretation of points shown on graph. Simple interpolation.	Letter as object or letter evaluated but there is a need to use mathematical syntax. What can you write for the perimeter of this shape?	Finding the area when not made up of half units or when one dimension must be found. Conservation of complementary volume and finding number of cubes in cuboid (cubes shown).	Can still only cope with two dimensions but is more critical, and so can recognise the inadequacy of some Stage 1 responses.

John leaves home to go to a disco in Cambridge 3 miles away. He walks one way takes a bus the other way. This is a graph of his journey.

At what time did he get to the disco?

What do you think he was doing between 7.00 and 8.45?

$p = \ldots$

$a + 3a$ can be written more simply as $\underline{4a}$.

Write this more simply where possible:

$2a + 5b + a =$

Volume of C ....

Shaded area =

Explain why C is not the centre of rotation.

Reflect the dot in m

Draw mirror line, if possible

Stage 3

The main difference between the items in Stage 2 and those in Stage 3 is the appearance of the first measure of abstraction − questions are not always tied to a diagram or the child is asked to hypothesise about situations which are not shown. The harder items require the use of a strategy, not a simple interpretation, with the result that many children show an incorrect method of solution such as the 'addition strategy' in the Mr. Short and Mr. Tall problem or make errors because they focus perhaps on one aspect of the question and ignore others, e.g. in Fractions the child has to cope with $\frac{2}{7} = \frac{\triangle}{14} = \frac{10}{\square}$ and see that $\frac{10}{\square}$ is connected to the $\frac{2}{7}$ as well as to $\frac{4}{14}$.

The large gap between the ability to interpret pictorial data and the understanding of an algebraic equation and its relationship to a graph means that there are no Graphs items at this stage. Nor are there any Measurement items. The Stage 4 items in Measurement require the child to adapt formula, e.g. for the area of a triangle and this is apparently much harder than Stage 2 requirements. Two levels of Ratio and two of Reflections and Rotations are in Stage 3. The Ratio items can be solved by a building up method − 'take it once, take it again, take a half and add' but they get progressively more difficult when the method involves a fraction other than one half or is essentially a counting down and not building up e.g.

```
                         25 cm                                    Z
─────────────────────────────────────────────────────────────────
                15 cm                          Y
──────────────────────────────────────────────
           10 cm              X
───────────────────────────────
```

Three other eels, X, Y and Z are fed with fishfingers, the length of the fishfinger depending on the length of the eel. If Z has a fishfinger 10 cm long, how long should the fishfinger given to X be?

Two percentage questions occur in this stage and are comparable in facility to the Decimal items dealing with multiplication by 100 or writing tenths as hundredths. The question $60 \div 0.3$ occurs in Stage 3, it is preceded on the test paper by $60 \div 3$ which is much easier. In $60 \div 0.3$ the answer is contrary to common sense i.e. sharing always makes it smaller.

In Fractions the questions require the use of equivalence but in each case the child has to apply the equivalence, e.g.

I am putting tiles on the floor; they are shown shaded. What fraction of the floor has been tiled?

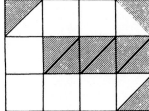

(Required answer $\frac{9}{24}$ or $\frac{3}{8}$)

In Algebra, Stage 3 items require the child to interpret and manipulate letters as unknown numbers rather than as objects, e.g.

$$r = s + t$$

$$r + s + t = 30$$

$$r =$$

The problem is no longer a simple collection of what is given.

In Vectors the questions deal with chains of moves when not all are shown on a diagram, e.g.

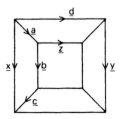

Which vector, or vectors, are equivalent to x̲?

The questions on the Positive and Negative Number paper involve subtraction, which is not easily modelled and tends to give answers contrary to what might be called common sense, e.g. ⁻2 − ⁻5 = + 3. The child who does not find this surprising has moved beyond the whole number maxim that subtraction always makes smaller.

In Reflections and Rotations Stage 3 questions involve reflections where the object, mirror line or image are not positioned on the page and rotations where the centre is not on the subject.

Stage 3 shows the first emergence of abstraction and the formulation of problem solving strategies (including incorrect strategies such as 'addition' in Ratio). Other items require decisions to be made on the size of the answer or the size of the measuring unit. The hardest Stage 3 items are successfully solved by 20—30 per cent of the sample. If mathematics teaching is designed to enable a child to face a new problem and invent a method of solution then the majority of secondary school children have evidently not reached this stage. Diagrams 5 and 6 illustrate the contents of Stage 3.

Diagram 3 Stage 3

Ratio	Fractions	Decimals	Vectors	Positive and negative nos
Must find rate or builds towards an answer (only needs to take half as much again). Operation on fraction.	Questions where more than one operation is required, e.g. equivalence followed by addition or subtraction.	Harder equivalence, relationship between adjacent and nonadjacent columns.	More advanced abstract moves. Maze question. Marking family of vectors successfully.	Subtraction $+5 - {}^-5 =$ ${}^-6 - {}^+3 =$
Onion soup recipe for 8 persons 8 onions 2 pints water 4 chicken soup cubes 2 dessertspoons butter ½ pint cream How much cream do I need for 6 people?	I am putting tiles on the floor. They are shown shaded. What fraction of the floor has been shaded?	Divide by one hundred $3.7 \to \ldots$ $60 \div 0.3 = \ldots$		
Mr. Short's height is shown measured in paper-clips. Mr. Short's height is 4 matchsticks. Mr. Tall's height is 6 matchsticks. How many paperclips are needed for Mr. Tall's height?			Mark any vector equivalent to a Mark any vector equivalent to c	

Diagram 5   Stage 3 (cont'd)

Graphs	Algebra	Measurement	Rotation and Reflection
No items	Able to work with letter as specific unknown, and to accept the lack of closure that this implies. $e + f = 8$ $e + f + g = \cdots$ What can you say about r if $r = s + t$ and $r + s + t = 30$	No items	Begins to be able to solve problems analytically. (e.g. can work with reflections that go off the page, and can transform object by breaking it down into distinct parts). Can also construct link between centre and object not on the centre. Reflect in m 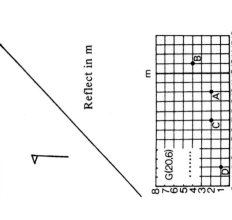 Rotate through a quarter turn (anticlockwise).

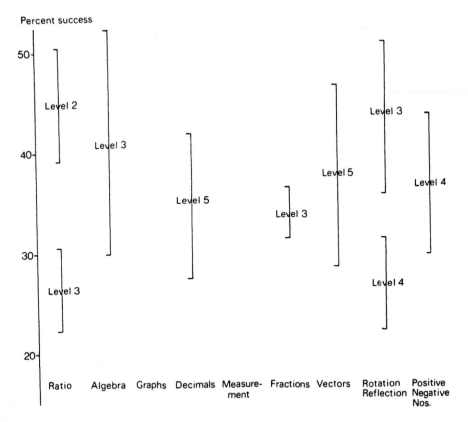

Percent success

Diagram 6   Stage 3 levels   Facilities from 3rd year sample.

Stage 4

Items in this last stage involve abstraction as well as the application of a fund of knowledge to the solution of problems. Many 'child methods' such as building up an answer by addition (in Ratio) are no longer sufficient and the child is faced with problems which require a different method. Such methods may have been taught but the children who now produce them when needed are probably those who could assimilate and store them when taught and who can now decide when to use them.

The line delineating Stage 4 is at a facility level of 30 per cent for 15 year olds, 20 per cent for 14 year olds and 18 per cent for 13 year olds, showing a steady increase of performance with age, the percentages themselves display the fact that the vast majority of children even at age 15 + are not able to deal with abstractions and application in mathematics. In Decimals and Fractions the questions require the child to appreciate the nature of these new numbers and not be firmly fixed within the set of whole numbers. He must be able to see that there is an answer to (16 ÷ 20) and not simply say 'the question is impossible because you cannot divide a small number by a larger', as he did when restricted to whole numbers. Stage 1 and 2 items in Fractions are very often of the type 'what name do we give to this', Stage 3 items require rather more, in that the connection between elements is needed. Stage 4 items need a departure from concrete referents, in particular they need a recognition that multiplication or division is needed and an ability to carry out that operation. The same need for abstraction appears in the Graph questions, although a diagram is presented (y = 2x) the child is no longer asked simply to describe what the diagram shows in terms of (1) a number (e.g. how many children came to school by train?) or (2) a pair of numbers (coordinates), but as a relationship between sets of unknown numbers. In Algebra, the child must not only work with unknown numbers, but must be able to coordinate operations, and relationships between sets of unknowns, and must be able to avoid using letters as objects in items that specifically involve objects. In Vectors the Stage 4 items deal with combining vectors or writing one vector in terms of others. Similarly in Geometry children have to combine reflections and rotations and interpret the results.

The two questions on volume require the use of a formula where the values are fractions or a formula which must be halved. On interview children often counted layers and did not use a formula which often leads to incorrect answers. In finding the volume of a triangular prism, the knowledge of the area of the base triangle which is not shown drawn on squared paper is required and although counting layers is feasible, the volume of each layer is far from obvious, e.g.

Find the volume of

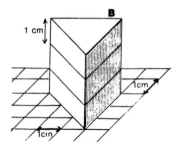

All the items which tested the knowledge of the infinite nature of numbers (e.g. How many numbers between 0.41 and 0.42?) occur at the end of Stage 4 or are of comparable difficulty. The fact that so many children choose a small finite number as the answer shows how far from truly understanding the range and power of mathematics most children are.

It is not until Stage 4 that mathematics as an abstract system and not merely a way of quantifying real world phenomena appears and even then it is usually the hardest Stage 4 items that embody this aspect. About 10 per cent of our child population seem to appreciate even the most elementary concepts in the type of mathematics which could be described as a system of abstractions.

Diagrams 7 and 8 illustrate the contents of Stage 4.

Diagram 7   Stage 4

Ratio	Fractions	Decimals	Vectors	Positive and negative nos
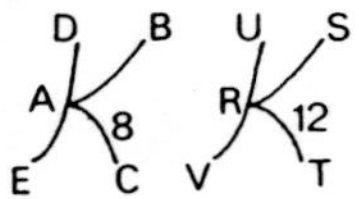 Recognises that a Ratio (multiplier) is needed. Multiplier non integer. (Hardest 2 items have added complexity in numbers or question.) 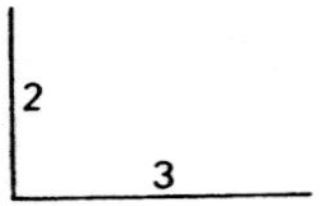	Division and multiplication of Fractions. a, b, d, are positive whole numbers $\frac{a}{b}$ is less than $\frac{a}{d}$ when 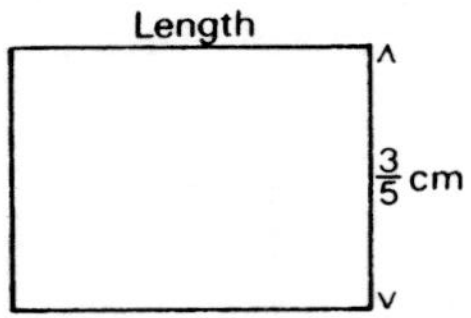	Decimals as a result of division. Divide 24 by 20. *How many different numbers could you write down which lie between 0.41 and 0.42*	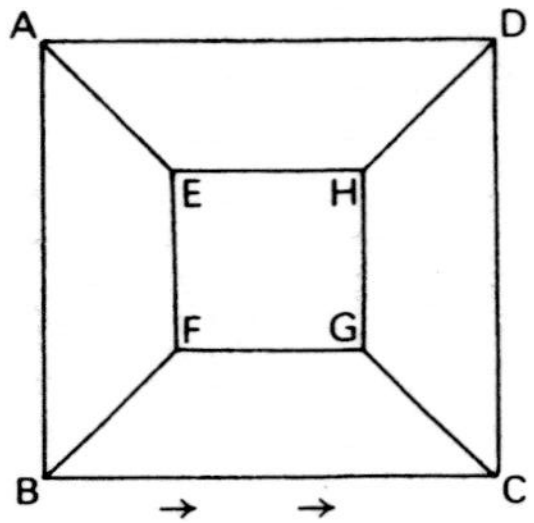 Interpretation and explanation of answers found. Commutativity of +. Use of Vector equivalence. Chains of Vectors with –ve signs required. 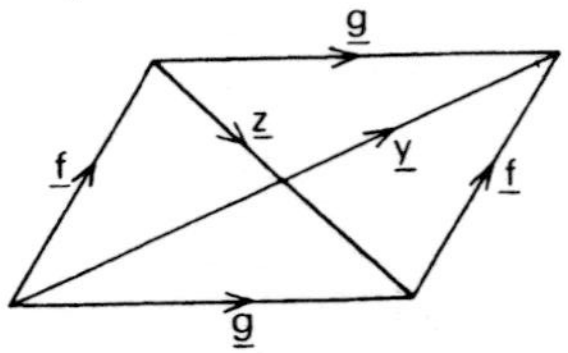	No items

2
3

enlarge with new base

5

Length

$\frac{3}{5}$ cm

Area = 1/3 sq. cm.

Length =

These two letters are the same shape, one is larger than the other. AC is 8 units. RT is 12 units. The curve AB is 9 units. How long is the curve RS?
[Drawn accurately on test.]

$\underline{y} =$
$\underline{z} =$

$\overrightarrow{AE} + \overrightarrow{HG} =$

Diagram 7   Stage 4 (cont'd)

Graphs	Algebra	Measurement	Reflection and rotation
Relation between Algebraic expression and graph	Manipulation of specific unknowns, where this involves a coordination of operations (in order to avoid ambiguities like $4 \times n + 5$) or where the letters represent numbers of objects, or their cost, etc., rather than the objects themselves. Use of letters as variables (2nd order relations).	Application of area or volume formulae in situations with half units or halving formulae.	Single rotations (of a quarter turn) with object and line from centre to object at any angle. Combinations of reflections and rotations.

Graphs

Find an x and y that make $x + y = 8$ true.

What is the equation of this line?

Algebra

Cakes cost c pence each and buns cost b pence each. If I buy 4 cakes and 3 buns, what does $4c + 3b$ stand for?

Which is larger, 2n or $n + 2$?

Explain

Measurement

Find the volume of

Find the volume of

Reflection and rotation

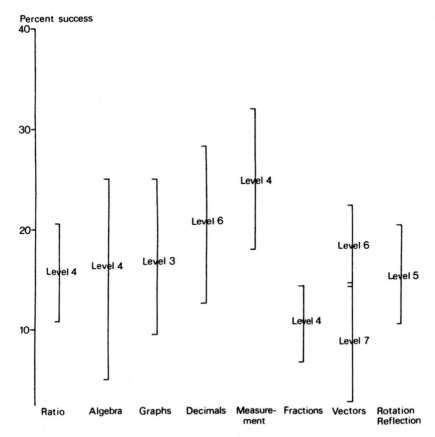

Diagram 8   Stage 4 levels   Facilities from 3rd year sample.

14   Implications for teaching

The research reported in this book was designed to investigate the formation of a hierarchy of understanding in mathematics. The resultant hierarchies were based both on the methods used by children during interviews and also the answers to written tests given to a large sample of English children. The interpretation placed on these results is necessarily opinion rather than fact, albeit informed opinion based on empirical data. This chapter on the implications for teaching mathematics is therefore rather different from those that precede it. The research was not based on classroom observation nor did we investigate the comparative merits of different textbooks so that though we have some information we cannot make pronouncements on how the children we tested were taught. We do have some information from teachers on what had been taught however.

Certain topics investigated, such as Reflections and Rotations, Matrices and Vectors can be classed as 'modern mathematics' in that they are relatively new to the secondary school syllabus. In both these chapters the authors state the amount of work done by the children on the topic varies considerably from school to school. One of the main aims of those who introduced 'modern mathematics' was the provision of a unified structure of mathematics so that children would see the same laws in operation within different topics. It was hoped that a child faced with a new mathematical system might ask 'What are the elements; which operations combine those elements; are the commutative, associative and distributive laws applicable etc?' This aim does *not* appear to have been fulfilled, e.g. children when dealing with fractions do not see the elements of the set in which they are working as different from the whole numbers since they apply the 'rules' of whole numbers. The laws of operation appear to be given little relevance by children, even when the elements are whole numbers. Between 20 per cent and 50 per cent of children in their first year in secondary school may not recognise the difference between a subtraction expression and its inversion, for division the figure lies between 45 per cent and 80 per cent, with only between 20 per cent and 30 per cent giving positive evidence of distinguishing the correct order consistently.

The evidence from the interviews illustrates this. A typical reply to the division items on the Number Operations paper was given by Ian (aged 12)

Ian —— 391 divided into 17 (indicates 391 ÷ 17) and this
one – 17 divided by 391 (indicates 17 ÷ 391).

The recognition of the commutative law and its significance in vectors is classed as an attribute of Level 6. In matrices it appeared to be only the children who had been taught the operation of multiplication who recognised such a law might exist, probably because the fact that multiplication was not commutative had been stressed.

Much of what appears in the hierarchies has been in the secondary school syllabus for many years and might be regarded as common to both modern and traditional mathematics. The individual items on the tests may not have been in a form that commonly appears in text books but they were designed to test understanding rather than 'rote learned' rules.

The results

The results of the CSMS research have far reaching implications for the teaching of mathematics at the secondary level. The overwhelming impression obtained is that mathematics is a *very difficult* subject for most children. We have shown that understanding improves only slightly as the child gets older and that there are children at every level in each of the secondary age groups investigated. Fifty per cent of our secondary child population can deal with the demand of Stages 1 and 2, that is they can cope with new terminology and first operations on elements e.g. addition of fractions. They can deal with problems that require only one or two steps for a solution but any demand for abstraction or even the formulation of a strategy for solution is beyond them. In the secondary school we tend to believe that the child has a fund of knowledge on which we can build the abstract structure of mathematics. The child may have an amount of knowledge but it is seldom as great as we expect. The spiral curriculum in which it is intended that we year by year build the fundamentals of mathematics may in practice have to be reduced to two dimensions and not three i.e. continual reiteration of the same points. The teacher may not in fact be building but *re-teaching* and so has to return time and again to what was thought to be already assimilated by the child. The re-teaching is very often seen as a revision exercise and does not take into account the necessary introduction of the topic. If one knows that the topic is new and is being presented for the first time one approaches it slowly. If on the other hand one thinks the child is in possession of certain facts then the approach is rather different. Gaps in the knowledge are recognised but the exercise is seen as one of 'patching up'. It seems likely that different embodiments, more concrete referents must be brought in all the time to illustrate further the base on which new mathematics is to be built. The building-up of mathematical knowledge can take place but it is a brick at a time and not a whole wall!

The type of mathematics which appears in the later Stage 3 and all of Stage 4 deals with abstraction, strategies for the solution to problems and a thorough familiarity with fractions and decimals and is very different from that in the first two stages. The questions which truly require abstraction belong in Stage 4, e.g. generalising fraction concepts ($a/b < a/d$ when?), stating the algebraic relationship symbolised by a graph. This is the type of mathematics recognised by the professional mathematician. The early stages may be regarded as a form of social Arithmetic or Mathematical literacy.

The results of the longitudinal studies reported in Chapter 12 show that achievement is closely linked to the IQ score of the child and that although each child progresses as it gets older the progression is not to the same point. Indeed it is perfectly logical that a child who is at Stage 0 at age 13 is not going to necessarily catch up with a 13 year old who starts at Stage 2. The type of mathematics given to the children must be tailored to their capabilities. It is *impossible* to present abstract mathematics to all types of children and expect them to get something out of it. It is much more likely that half the class will ignore what is being said because the base on which the abstraction can be built does not exist. The mathematics must be matched to each individual and teaching a mixed ability class as an entity is therefore unprofitable. Indeed, in a class where the children are supposedly matched on attainment there will still be a wide range of individual needs.

Primary mathematics

The primary school mathematics curriculum includes operations on whole numbers, graphs of pictorial data and measurement in which a unit is repeated and the answer can be obtained by counting. All these topics are within the range of understanding of 80–90 per cent of the secondary school children. If one however considers the amount of *time* spent in the primary school on these aspects of mathematics one realises that the acquisition of knowledge is a slow process. Furthermore it is likely that apparatus, concrete referents and 'aids' to learning were much in evidence in the primary school, certainly in the early years. However, a recent survey carried out by the Schools Council (M. Ward, 1979) shows that at the age of ten many schools feel the child should not be using mathematical apparatus. In this survey the replies to the question 'about how often does a child use maths apparatus?' are shown in Table 14.1.

Table 14.1 Use of apparatus by ten year olds — Ward 1979 (percentage)

No Data	9
Never	1
A few times a month	25
Once or twice a week	45
Most days	20
Every day	0

There is research evidence to show that certainly primary school children benefit from the use of concrete materials (Ekman 1967 and Punn 1974 working with 9–10 year olds). For some secondary school topics it is difficult to think of apparatus that would be relevant but certainly all those topics which have traditionally been introduced in the primary school and which we have seen need to be retaught to many children in the secondary school should benefit from the use of apparatus. These topics are usually extensions, e.g. whole numbers into decimals and fractions, measurement into the use of formula etc. There appears to be a belief on the part of both children and teachers to think of apparatus as something rather sinful, e.g. 'You surely don't need to use bricks for doing that!'

The child usually feels secure when working with whole numbers (albeit small integers and simple operations) but he often applies what he thinks are rules for whole numbers to decimals and fractions. The power of these new numbers has *not* been conveyed to him. The fact that they can now find an answer to '16 ÷ 20' is significant but unrealised by most children. On interviewing children we found many who also avoided using a fraction at all cost. Some avoided multiplication of whole numbers. The avoidance of multiplication by using repeated addition means that for these children '⅓ × ¾' has no meaning and it is not surprising that items of this type belong in Stage 4.

Most of mathematics is concerned with understanding and ideas rather than skills but the ability to use a ruler to measure a line segment might be regarded as a skill. The assumption that every child in a secondary school class can use a ruler accurately is fallacious, many children do not count the units contained in the line segment but the end points of those units. The same assumption applies to the knowledge of the standard algorithms of addition and subtraction, the few examples requiring such computation which appeared on the CSMS test papers, had no apparent increase in facility commensurate with age. For example about 60 per cent of every year could compute 2312 . It seems likely that if the child could not

$$\begin{array}{r} 2312 \\ -\ 547 \\ \hline \\ \hline \end{array}$$

do this type of 'sum' when he entered the secondary school he was unlikely ever to be able to do it. The teacher should be aware of this apparent gap in the child's knowledge and deal with it accordingly.

We also assume that if a child can 'do a sum', he is also able to *choose* which operation to use when faced with a number problem. The chapter on Number Operations showed that this was far from true for many children. For every child like Tony there is a Tracey!

Tony (aged 13)   12 (very quick response)
Interviewer   What did you do with the four and the three?
T   Four different sorts and three different sizes, four times three or three times four.

Tracey for 84 × 28.
If Tom has 84 sweets and Susan had 28 if they timsed it how many would they have.

Traditionally we teach the operation of division last, possibly because we are thinking of the rules for long division which are difficult to explain and because within the set of whole numbers division is limited. The research on the actual *recognition* of appropriate operations (Number Operations and Decimals) shows that the order of difficulty is more likely to be: addition, subtraction, division, multiplication.

Algorithms

The vexed question of when and whether to teach rules or algorithms constantly arises. The interviews on every topic showed that the children for the most part did *not* use teacher taught algorithms. Some rules however did appear as part of their repertoire, e.g. addition of whole numbers (although it was highly likely that the child could not explain what he was doing). To a great extent children adapt the algorithms they are taught or replace them by their own methods, it is only when these methods fail them that they see a need for a rule at all. We appear to teach algorithms too soon, illustrate their use with simple examples (which the child knows he can do another way) and assume once taught they are remembered. We have ample proof that they are not remembered or sometimes remembered in a form that was never taught, e.g. to add two fractions, add the tops and add the bottoms.

The teaching of algorithms when the child does not understand may be positively harmful in that what the child sees the teacher doing is 'magic' and entirely divorced from problem solving. The fraction test paper provided an illustration of this, since both computations and problems were presented. Often the problem was easier than the computation, showing that the children did not necessarily use the computational method to solve the problem. One purpose of teaching mathematics in school is to provide children with the skills and information to solve problems. It is to this end that we teach them rules which we as adults think are short cuts to solutions of problems. If to the child the algorithms are not based on reality and to a large extent are seen as something separate and not necessarily logical, then we are certainly not fulfilling the purpose for which the teaching was intended. In order to avoid teaching rules which the child cannot apply we must first discover what the question we are asking him to answer means to *him*. If he does not see it as 'multiply' then giving him an algorithm for multiplication is not apposite. Secondly we must find out what method he normally uses to solve the problems of this type and build on that.

Some methods used by children are very idiosyncratic but correct, the more idiosyncratic of couse, the less the connection between what the teacher says and what the child does, e.g. Tim (aged 13)

$$\begin{array}{r} 38 \\ + 27 \\ \hline \\ \hline \end{array}$$

T   When I first did it I thought I'd try doing it in my mind — I said 3 and 2 — 30 add 20 is 50, so I went to add 7 and 8 together. Then I said, 8, 9, 10, take 2 off the 7 is 5, and that makes another one, that's 6. You're supposed to carry but I don't always do that.

$$\begin{array}{r} 51 \\ - 28 \\ \hline \\ \hline \end{array}$$

T   2 from 5 is 30, then I said 8 take 1 leaves you 7, then I said 7 from 30
leaves 23.
(Tim then seemed surprised when told that this was not the way most
people did it).

In previous chapters child methods which appear to be in general use, in that a
number of children from different schools seem to use them, have been described.
These should be particularly borne in mind when a topic is being taught.

In most classrooms considerable verbal communication takes place but it is
seldom the teacher with one child, except in the context of the teacher 'telling'
the child. If one is to assess the level of understanding of the child and why he is
making mistakes there seems to be no alternative but to talk to him. With a class
of 25 children this may seem impossible but half an hour a day for five weeks
would provide some clue to each child's understanding. Finding the time is
difficult but some primary schools have instituted this arrangement by time-tabling
two classes together at some times. Team teaching is usually viewed as two or three
classes together with two or three teachers. If the arrangement were extended to
include one of the teachers spending his time assessing and talking to an individual
child whilst the others coped with the rest, then valuable assessment could take
place. We tend often to assess the progress of a child by stating what he does that is
correct and what he does that is incorrect rather than asking ourselves why he is
correct or why he is wrong.

Talking to individual children may soon clear up certain misconceptions which
the child will not voice in public. It was noticeable during the interviews, carried
out as part of this research, that children were learning simply by voicing their
thoughts.

Language

The language used in mathematics lessons is often technical, sometimes needlessly
so. It was apparent when interviewing fourteen year olds that the words 'perimeter'
and 'area' were not part of their normal vocabulary and had to be redefined. Other
words may have taken-on a limited (and incorrect) meaning. Examples from an
interview with Faith (aged 14) illustrate two of these misconceptions:

Interviewer: 10 sweets are *shared* between two boys so that one has 4
more than the other. How many does each get?

Faith: That's wrong, if you *share* they each have 5, one can't have
4 more.

2. Interviewer: I think of a whole number, add one to it. Can the result be
divided exactly by two?

Faith: If you had a whole number and added 1, it would be odd.
It won't work.

I   Will it work anywhere?

F   No. adding on 1 makes it odd, even if its thousands.

I   How about starting with 5.

F   I thought *whole* number meant *even.*

In the chapter on Ratio, mention has been made of the conflict between the everyday use and the mathematical sense of the word 'similar'. 'Bigger' is another word which is used frequently and often ambiguously, e.g. in the context of the comparison between two cubes (A side 1 cm, B side 2 cm) we have three alternative meanings:

> B is 8 times as big as A (volume)
> B is twice as big as A (edge measurements)
> B is 4 times as big as A (surface area).

Another term often used incorrectly is 'straight'; for many children a slanting line cannot be straight because it is not perpendicular to the edge of the page.

Errors

We have shown that there are certain errors made by children from many different schools. If these errors are as widespread as seems likely then when teachers present a topic they should be aware of what is likely to occur. They may even be able to build into their presentation examples which show the illogical outcome of the incorrect method. Instead of always asking a child to do a series of very nearly identical problems a useful exercise is to present a problem done in different (and erroneous) ways and ask the child to state which are wrong answers obtained from which wrong methods. To correct or discuss homework or set exercises simply by repeating the 'teacher method' seems to be of limited value. If that is all that is needed then the child would have been able to take in the method on its first presentation. If one corrects a ratio problem by writing $\frac{x}{a} = \frac{y}{b}$ and substituting for the letters, one is not correcting what the children did. 'A' may have been building up to an answer by 'taking once, taking twice, take a half', 'B' may view all enlargement as doubling. Few of the class will have made errors in the computation of $\frac{x}{a} = \frac{y}{b}$ because few will have *used* it. The method many of us use is that of 'showing' one child at a time 'how to do it'. What we mean here is how we would have done it, not how the child would have because we never ask him. Very often in this latter case we end up by doing the question for him and then set him some more identical questions in which he is to repeat exactly what we have just written. Perhaps we should get away from 'I'll show you' and into 'let us discuss what this means'.

If children could recognise that they have made an error then they themselves might do something about correcting it. Many are just too thankful to have an answer, any answer, to even dare investigate further. It may be that only few children are capable of recognising an error and learning from it, as seems to be suggested by the research of Ramharten and Johnson (1949) with twelve year olds which involved the testing of subtraction of fractions. After the test the students were given a guide sheet with the examples of the test worked correctly. Each pupil was told to study the guide sheet (thinking aloud) in preparation for retesting. The comments of the pupils were recorded and it was found that those who were 'good' achievers on the original test exhibited greater insight and greater disgust concerning their previous errors and experienced other kinds of help from the guide sheet more often than the 'poor' achievers. Further research on this whole topic is needed.

As teachers we possibly do not put enought recognition of errors into our mathematics teaching. When teaching similarity for example do we present a set of similar figures or do we ask 'which of these look the same'? Is the recognition aspect introduced and then never reinforced? Does a child see that what he has drawn is grossly dissimilar to the one with which it is to correspond? The fact that children do not recognise the non-similarity of two figures supposedly the same shape has been commented on in the context of both matrices and ratio. Very often we say to children 'always check your answer', do we mean a) look at it, b) is it the right order of magnitude c) do it all over again and see whether you have the same answer or d) think of a different operation which applied to what you have should give you what you started with. This last is surely what we should be aiming for. The fact that children show very little flexibility in moving from one operation to another is demonstrated by two adjacent questions on the Fractions Computation paper:

$$3 \times 10\tfrac{1}{2} \text{ [about 76–84 per cent success]}$$
$$40 \div 10\tfrac{1}{2} \text{ [about 20–27 per cent success, accepting}$$
$$3 \text{ rem. } 8\tfrac{1}{2} \text{ as an answer]}.$$

To the children these two questions were asking entirely different things and their first thought was probably 'what do you do when it's division?'

Another aspect of being critical about the solution one has obtained to a problem is that of rigour. On the CSMS papers we did not ask children to prove but we did ask 'why?' The replies to this were often tautologous, the child essentially saying it is so because it is so. Sometimes the replies were no more than an expression of feelings, as on the Graphs test paper when asked if two lines were parallel and why, many children stated 'it looks like it'. We possibly as teachers often accept answers of this sort in class, in an effort to be encouraging we say 'yes' and go on to a child who might give us a more cogent reason. Does the first child actually see any difference in quality between what he has said and what his friend has provided as a reason? If we do nothing to make him aware of what is a valid explanation and what is not we are certainly not fulfilling the role of mathematics in that it 'trains people to think'.

If a child can successfully deal with one aspect of mathematics the next stage may not be the one we as teachers see as a logical next step. Any complexity added makes the problem very much harder, not just a little harder (notice the large gap between Stage 2 Graphs and the next stage or Stage 1 Ratio and the next stage). Note too the drop in facility when larger numbers are introduced into the recognition of operations – they are much more difficult to even think about let alone do anything with. Recognising the same operations when the elements are decimals is more difficult still.

Addition appears as an incorrect method in many of the topics because the child in doubt as to which operation should be used usually adds, e.g. Maria (aged 13) used 'you add them' as a universal strategy for any word problem, although she did at least pause at the end to reflect 'That's funny – how come they're all adds?' In Ratio adding in order to enlarge was very prevalent, e.g. Debbie (aged 14) faced with an enlargement of 6 to 9 said 'If it was double it would have been 12. So its only 9, they added 3 on. In doubling its the same number again . . . You can't have six times something to give 9. You can *plus* it.'

Here doubling is seen as an additive (and correct) operation. Multiplication was often carried out by using repeated addition which is usually sufficient for whole numbers but no use for fractions. The operation of multiplication itself was much more difficult than one might expect, both in terminology used ('times' in Number Operations) and in recognition of its use.

Fractions and operations upon them, are not a simple step from whole numbers, except for 'one half' which seems to be an honorary whole number. Whenever a fraction was present in a problem, the facility dropped (see the cream question in Ratio when the introduction of a fraction meant the facility dropped 50 per cent). Multiplication of fractions in Measurement, Fractions and Ratio was always in Stage 4.

Diagrams appeared to be useful in Vectors and Fractions, pictures corresponding to decimal notation seemed to help on the Decimal paper. Stage 1 items were very often accompanied by a picture. The only case in which they seemed to provide a distraction was in time and distance graphs when the essential relationship time/distance was lost in the visual aspects of the graph. Many children looked at the picture and described it in terms of going up or left rather than stating the meaning of the line segments. In graphs, the teacher should also be aware of the danger of always presenting the axes in the same way, e.g. height is always the axis that goes up the page. Many children described Helen (H) as tall and thin in the following graph interpretation

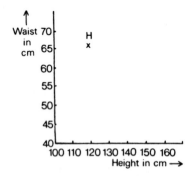

Mark in John (J) whose height is 140 cm and waist 60 cm.

What can you say about Helen's appearance?

It is hoped that the presentation of a hierarchy in each topic in this book will give teachers some guidance on the sequencing of the topic they teach, it might for example be more rewarding for both teacher and pupil if instead of pursuing one topic relentlessly, they side step to a comparable level in a different topic. Very often secondary school teachers are reteaching what was first presented in the primary school. In many cases it was not understood when first presented, or even when 'taught' the second time. After a history of failure at decimals or fractions it is little wonder that the next teaching of it in the secondary school is greeted with dismay. Is there really any point in teaching something we know most children will not understand? One reason given for doing this is that the child will become familiar with the idea and understand it later. We have no proof of this, in fact our results show that the understanding does not 'come'. Surely all that happens is that the child becomes familiar with a lack of success and that mathematics is something you do but it makes no sense. Is the possible answer to state that fractions and decimals are topics fit only for secondary school children and so encourage the primary school to limit its number work to whole numbers? This does not mean an encouragement to practice computation *only* for five years but instead to explore the problem-solving situations in which whole numbers can be used. This does not mean that the brighter child is penalised. In order to stretch him we do not need to present him with a new set of numbers but with new situations in which he can use the numbers he knows how to handle.

All children make some progress but it is very *slow*. The less able pupil may be slowly moving though the type of mathematics which occurs in stages 1 and 2 and if he is given the type of mathematics which is essentially abstract, he is unlikely to assimilate any of it until he is 'ready'. The problem is always to find what his 'knowledge base' is and build on that. As teachers we have expectations of what a child 'should' know, very often based on intuition and usually very different from the actuality.

Most of this book has been a description of the results of testing children with the CSMS tests. Teachers may wish to use the tests themselves either to assess the level of a child's understanding before they introduce a new topic or after they think they have taught it. Each test is available (with a Teacher's Guide) from the National Foundation for Educational Research (NFER).

Appendix 1

Schools used

School
Number

01	Mixed Comprehensive (Bristol)	29	Middle School (Sheffield)
02	Mixed Comprehensive (Bristol)	30	Middle School (Sheffield)
03	Mixed Comprehensive (Bristol)	31	Mixed Comprehensive (Sheffield)
04	Mixed Comprehensive (Herts.)	32	Middle School (Sheffield)
05	Boys Comprehensive (Herts.)	33	Mixed Comprehensive (Sheffield)
06	Mixed Comprehensive (Herts.)	34	Boys Comprehensive (Kent)
07	Boys Comprehensive (Herts.)	35	Mixed Comprehensive (Kent)
08	Mixed Comprehensive (Herts.)	36	Boys Comprehensive (Kent)
09	Mixed Comprehensive (Herts.)	37	Girls Comprehensive (Kent)
10	Mixed Comprehensive (Herts.)	38	Mixed Comprehensive (Kent)
11	Mixed Comprehensive (Herts.)	39	Mixed Comprehensive (Leicester)
12	Mixed Comprehensive (Herts.)	40	Mixed Comprehensive (Essex)
13	Mixed Comprehensive (Glos.)	41	Boys Comprehensive (Coventry)
14	Mixed Comprehensive (Glos.)	42	Boys Comprehensive (S. London)
15	Girls Grammar (S. London)	43	Mixed Secondary Modern (Suffolk)
16	Mixed Comprehensive (Coventry)	44	Girls Comprehensive (S. London)
17	Girls Grammar (Plymouth)	45	Mixed Comprehensive (Manchester)
18	Mixed Comprehensive (Notts.)	46	Mixed Comprehensive (London)
19	Mixed Comprehensive (Notts.)	47	Mixed Comprehensive (S. London)
20	Middle School (Leeds)	48	Mixed Primary (Edinburgh)
21	Girls Grammar (Private) (S. London)	49	Boys Comprehensive (London)
22	Girls Grammar (Herts.)	50	Mixed Comprehensive (Humberside)
23	Boys Grammar (Private) (London)	51	Middle School (Surrey)
24	Mixed Comprehensive (Salford)	52	Mixed Comprehensive (Middx.)
25	Girls Comprehensive (Leeds)	53	Middle School (Surrey)
26	Mixed Comprehensive (Somerset)	54	Mixed Comprehensive (London)
27	Mixed Comprehensive (Somerset)		
28	Mixed Comprehensive (Somerset)		

Sample breakdown 1976, 1977

Test	Year	School	No. of cases		Test	Year	School	No. of cases	
Algebra '76	2	5	182		Vectors '76	2	20	52	
2923		6	220		2210		21	68	120
		7	82			3	3	107	
		8	123				4	142	
		15	86				5	163	
		16	276				6	202	
		20	133				8	121	
		23	26	1128			10	103	
	3	4	140				15	61	
		5	160				18	220	
		6	202				23	21	1140
		8	133			4	1	22	
		17	92				3	103	
		18	212				6	191	
		23	22	961			9	88	
	4	1	21				12	124	
		6	92				13	80	
		9	98				16	224	
		11	134				17	95	
		14	142				23	23	950
		16	220		Reflections	2	9	24	
		23	24	731	and		11	23	
	5	6	45		Rotations '77		12	38	
		15	58	103	1026		20	23	
							21	14	
							25	30	
Ratio '76	2	1	30				26	56	
2257		5	177				31	58	
		7	79				33	27	293
		8	119			3	7	19	
		16	277				8	64	
		17	89				14	50	
		20	21				15	15	
		22	8	800			16	144	
	3	4	141				17	33	
		8	121				18	48	
		10	106				24	76	449
		17	92			4	4	37	
		18	190				12	37	
		19	117	767			13	51	
	4	1	22				25	29	
		6	86				28	83	
		11	126				38	47	284
		14	137		Number	−1	20	102	102
		15	66		Operations	0	20	118	
		16	228		'75		22	32	
		22	25	690	976		38	110	

Test	Year	School	No. of cases	
Number Opn. contd.		39	25	
		41	38	
		44	147	
		48	27	497
	1	20	113	
		40	31	
		43	38	
		49	65	247
	2	45	87	
		50	43	130
Graphs '76 1396	2	1	26	
		2	9	
		3	96	
		6	189	
		17	60	
		20	9	389
	3	1	85	
		3	112	
		10	101	
		18	206	
		19	80	
		22	10	594
	4	1	15	
		3	107	
		9	98	
		12	118	
		13	75	413
Matrices M '77 535	3	9	63	
		18	40	
		21	67	170
	4	34	101	
		35	139	
		36	54	
		37	71	365
Matrices A '77 257	3	5	66	
		9	12	
		18	57	135
	4	34	52	
		36	48	
		37	22	122
Fractions 1 & 2, '77 555	1	17	22	
		21	70	
		29	25	

Test	Year	School	No. of cases	
Fractions contd.		30	26	
		32	48	
		34	55	246
	2	9	23	
		11	26	
		12	37	
		20	29	
		21	17	
		25	30	
		26	53	
		31	70	
		33	24	309
Fractions 3 & 4, '77 523	3	7	19	
		14	80	
		15	16	
		16	68	
		17	30	
		18	95	308
	4	4	35	
		12	32	
		13	23	
		25	23	
		28	46	
		38	56	215
Measurement '77 986	1	17	21	
		29	27	
		30	28	
		32	42	
		34	51	169
	2	9	51	
		11	57	
		12	77	
		20	52	
		21	26	
		25	27	
		26	102	
		33	52	444
	3	7	21	
		8	56	
		14	43	
		15	21	
		16	74	
		17	37	
		18	47	
		24	74	373

Test	Year	School	No. of cases		Test	Year	School	No. of cases	
Decimals	1	17	21		Positive	2	9	25	
'77		29	27		& Negative		11	34	
950		30	25		Numbers '77		12	34	
		32	44		818		20	31	
		34	53	170			25	60	
	2	9	23				26	56	
		11	28				31	70	
		12	39				33	24	334
		20	30			3	4	37	
		25	29				7	21	
		26	58				8	31	
		31	64				14	49	
		33	23	294			15	22	
	3	7	18				16	71	
		8	25				17	34	
		14	48				24	37	302
		15	19			4	12	35	
		16	74				13	24	
		17	31				25	27	
		24	32	247			28	49	
	4	4	35				38	47	182
		12	35						
		13	22						
		21	15						
		25	29						
		28	48						
		38	55	239					

Appendix 2

Item-item measures of association

Consider the fourfold table classifying individuals according to their performance on two test items:

Item 2

		Fail	Pass
Item 1	Pass	a	b
	Fail	c	d

a = number of individuals who pass item 1 and fail item 2 etc.

Then in this notation

$$\phi = \frac{bc - ad}{((a + b)(c + d)(a + c)(b + d))^{1/2}}$$

(ϕ is a special case of the product-moment correlation for dichotomous data)

$$H = \frac{bc - ad}{(b + d)(c + d)} = \frac{\phi}{\phi_{max}} \quad \text{(Loevinger, 1947)}$$

where ϕ_{max} is the maximum value of ϕ given the perceived difficulties of the items. See, for example, J. P. Guilford, *Psychometric Methods* (1954).
(Note: In the calculation of H item 2 must be the harder of the two items, i.e. $d < a$)

Kruskal's Gamma

This is a measure of association between two tests.
Two individuals scoring (X_1, Y_1) and (X_2, Y_2) on the two tests are said to be

concordant if	$X_1 < X_2$	and	$Y_1 < Y_2$
or	$X_1 > X_2$	and	$Y_1 > Y_2$
and *discordant* if	$X_1 < X_2$	and	$Y_1 > Y_2$
or	$X_1 > X_2$	and	$Y_1 < Y_2$

If P = number of concordant pairs, and Q = number of discordant pairs then:

$$\text{gamma} = \frac{P-Q}{P+Q}$$

See Goodman L. A. and Kruskal W. H., Measures of association for cross classifications, *Journal of the American Statistical Association* (1954).

Kruskal's gamma test/test

	Algebra	Ratio	Graphs	Fractions 1, 2	Fractions 3, 4	Measurement	Decimals	Pos. Neg. Nos
Algebra								
Ratio	.763							
Graphs	.788	.790						
Fractions 1.2	.740	.839						
Fractions 3, 4	.778	.853	.739					
Measurement	.709	.790	.795	.629	.664			
Decimals	.765	.800	.733	.810	.802	.605		
Pos. Neg. Nos	.747	.596					.797	
Reflections/ Rotations	.681	.591	.706			.681		.686

Values of the phi coefficient in each hierarchy

Test	Level	Mean ϕ	Median ϕ	Test	Level	Mean ϕ	Median ϕ
Ratio '76	1	.31	.17	Positive	1	.47	.44
	2	.39	.35	& Negative	2	.43	.40
	3	.41	.37	Numbers	3	.40	.38
	4	.41	.38		4	.38	.38
Number	1	.16	.16	Matrices	1	.41	.37
Operations	2	.31	.31	Core	2	.48	.47
	3	.38	.37		3	.44	.41
Decimals &	1	.32	.29		4	.40	.34
Place Value	2	.36	.35	Vectors	1	.37	.37
	3	.49	.48		2	.45	.42
	4	.40	.39		3	.44	.42
	5	.46	.45		4	.45	.44
	6	.43	.41		5	.44	.42
Fractions	1	.42	.44		6	.36	.33
1 & 2	2	.46	.46		7	.31	.30
(problems)	3	.45	.43	Graphs	1	.17	.16
	4	.40	.40		·2	.31	.29
Fractions	1	.49	.46		3	.41	.36
3 & 4	2	.44	.43	Measurement	1	.38	.36
(problems)	3	.44	.47	'77	2	.40	.34
	4	.34	.33		3	.44	.44
Algebra	1	.28	.28		4	.41	.33
	2	.42	.38				
	3	.44	.43				
	4	.40	.40				
Rotations	1	.36	.33				
&	2	.30	.32				
Reflections	3	.42	.37				
	4	.33	.33				
	5	.37	.36				

References

Calvert, B., Non-Verbal Test DH (NFER-Nelson, Windsor)

Collis, K.F., *A Study of Concrete and Formal Operations in School Mathematics: A Piagetian Viewpoint*, ACER Research Series 95 (Australian Council for Educational Research, Victoria, Australia, 1975a)

Collis, K.F., *Cognitive Development and Mathematics Learning*, PMEW Paper (Chelsea College, University of London, 1975b)

Collis, K.F., *The Development of Formal Reasoning*, report of an SSRC sponsored project carried out at the University of Nottingham during 1974 (University of Newcastle, New South Wales, Australia, 1975c)

Collis, K.F., 'Operational Thinking in Elementary Mathematics', in Keats, J.A., Collis, K.F. and Halford, G.S. (Eds), *Cognitive Development* (John Wiley, 1978)

Ekman, L.G., 'A Comparison of the Effectiveness of Different Approaches to the Teaching of Addition and Subtraction Algorithms in the Third Grade', Vols 1 and 2 (University of Minnesota, 1966), *Dissertation Abstracts*, 27A (1967) 2275−6

Goodman, L.A. and Kruskal, W.H., 'Measures of association for cross classifications', *Journal of the American Statistical Association*, 49 (1954) 732−64

Guilford, J.P., Psychometric Methods (McGraw-Hill, 1954)

Halford, G.S., 'An approach to the definition of cognitive developmental stages in school mathematics', *British Journal of Educational Psychology*, 48 (1978) 298−314

Inhelder, B. et al, *Learning and the Development of Cognition* (Routledge, 1974)

Karplus, R., Karplus, E., Formisano, M. and Paulsen, A−C., 'Proportional Reasoning and Control of Variables in Seven Countries', Advancing Education Through Science Orientated Programs, *Report ID-65*, June 1975.

Kieren, C., *Children's Operational Thinking Within the Context of Bracketing and the Order of Operations*, Paper presented at annual conference of IGPME, Warwick, July 1979 (Concordia University, Montreal, 1979)

Kuntz, B. and Karplus, R., *Intellectual Development Beyond Elementary School VII: Teaching for Proportional Reasoning*, paper given to the Psychology of Mathematics Education Workshop, Chelsea College, University of London, 1977

Loevinger, J., 'A Systematic Approach to the Construction and Evaluation of Tests of Ability', *Psychological Monographs*, 61 (1947) 4

Piaget, J., Grize, J.B., Szeminska, A. and Bang, V., *Epistemologie et Psychologie de la Fonction* (Presses Universitaires de France, Paris, 1968)

Piaget, J., Inhelder, B. and Szeminska, A., *The Child's Conception of Geometry* (Routledge and Kegan Paul, 1960)

Punn, A.K., 'The Effects of Using Three Modes of Representation in Teaching Multiplication Facts on the Achievement and Attitudes of Third Grade Pupils' University of Denver, 1973), *Dissertation Abstracts International*, 34A (May 1974) 6954−5

Ramharten, H.K. and Johnson, H.C., 'Methods of Attack used by 'Good' and 'Poor' Achievers in attempting to correct Errors in Six Types of Subtraction Involving Fractions', *Journal of Educational Research*, 42 (April 1949) 586—97

Renner, J.W. and Paske, W., 'Quantitative Competencies of College Students', *J. Col. Sci. Tech.*, May 1977

Sheppard, J.L., 'Concrete operational thought and developmental aspects of solutions to a task based on a mathematical 3-group', *Developmental Psychology*, 10 (1974) 116—23

Siddons, A.W., The Teaching of Elementary Mathematics (CUP, 1933)

Torgerson, W.S., *Theory and Mathematics of Scaling* (Wiley, New York, 1958)

Ward, M., *Mathematics and the 10 year old*, Schools Council Working Paper 61, (Evans/Methuen Educational, 1979)

Index